Mistaken for Granite:
earth science for rock watchers

Paperback version

Peter Macinnis

Polymoth Books
https://tinyurl.com/polymoth

Publication data

Author: Macinnis P. (Peter) (1944 –).

Title: *Mistaken for Granite*.

ASIN (e-book edition): B085BGM95Z

ASIN (paperback edition): B0858W4YSK

ISBN for print version: 9798620093632

Notes: Originally an e-book with internal navigation. Contains illustrations.

Target audience: curious minds from mid-teens to mid-90s.

Subjects: Rocks, geology, environment, climate change earth science.

Copyright: text and photographs copyright © Peter Macinnis 2020, line art either by the author or public domain.

Published by Polymoth Books, https://tinyurl.com/polymoth.

Cover illustration: Granite hills, Freycinet Peninsula. Tasmania. [Peter Macinnis].

Peter Macinnis is an award-winning Australian writer for both adults and children. His awards come from the Children's Book Council of Australia, the West Australian Premier, the Wilderness Society, The Educational Publishing Awards Australia (EPAA) and the Royal Zoological Society of New South Wales, among others. There is a full list of his awards at http://members.ozemail.com.au/~macinnis/writing/awards.htm.

Trained as a biologist, he cares about natural history, social history and the stories behind things, and so he has become well-regarded as an historian. He also talks on ABC Radio national from time to time, sometimes teaches adults how to do extreme research and data handling, and thoroughly enjoys being the visiting scientist at his local K-6 school.

Dedication:
For Christine, who has helped me bother many rocks.

Blurb

This is a book for people who wonder why rocks are as they are. Here, among other everyday things, you will find tales of wells that go (slightly) uphill; rivers that flow underground; ancient recycling; rotten rocks; pretty rocks; how and when rocks were made; waves and beaches; rocks that float, fall, bounce, bend and tip over; asteroids; hoodoos; Liesegang rings; falling things and weighing the planet; plate tectonics; crystals; lightning as a destroyer of rocks; glaciers; mud; sand dunes; caves; tombs; soil; and where to find a fortune in gold.

Those are the stories the rocks can tell, if you know how to read them. The rocks won't tell you (but this book does) about poets, playwrights and plagiarists; mad (maybe) and devious (certainly) scientists; altitude sickness; ringing bells in Boston; walking on and inside volcanoes; elephants in stiletto heels; golf in space; rocks in exotic locations; a tourist authority conspiracy; a quiz show that got it wrong; the art of making aqueducts; finding water in a desert; poison wells; fat strippers and oil wells; hot spots; fake fossils; pretending to be a wizard in Coimbra in Portugal (where the undergraduates wear Harry Potter cloaks); how (and why) the author smuggled a fossil; stone fortifications, monuments, bridges and buildings; rock inscriptions and art, and what they tell us; behaving oddly in art galleries; mapping the planet's surface and interior; gravity and finding exoplanets; telling the truth about cholera and lies about SARS; why climate matters and how we (and the world) will probably end, possibly sooner than we think. There are also brief references to dragons and pixies, but only in a soundly geological context.

A short foreword

> You cannot hope to bribe or twist,
> thank God! the British journalist.
> But, seeing what the man will do
> unbribed, there's no occasion to.
> —Humbert Wolfe (1885 – 1940).

This book is written with the intelligent reader in mind, somebody living in an age when that disease has smitten many of the world's scriveners, an age when, unless we care about the earth sciences, our grandchildren have no future.

Wolfe's epigram, written in the year he died, reminds us that the tricks we see in Fake News and Alternative Facts have been around longer than we realise. The techniques used by the gutter-press and the gutter-politicos have deep roots.

That person who looks you in the eye while picking your pocket and stabbing you in the back is just the latest to follow an old tradition. We need earth science, as never before.

Chapters

Chapters	iv
A preamble about a scramble.	1
Background to earth science.	10
1: How the earth was made.	34
2: Deep rocks.	68
3: How rocks wear away	75
4: Spicks and specks.	105
5: Rocks in beds.	113
6: Changing rocks.	142
7: Things going wrong.	150
8: Rocks that ignore the rules.	172
9: Water and geology.	181
10: Rocks that were once alive.	194
11. Making money from rocks.	221
12: Thinking about rocks.	233
13. Why dropping rocks on toes hurts.	253
14. Oh dear, is that the time?	263
15. End-of-the-earth science.	274
16. The end.	287
17. Glossary.	295
About the author	302

Note: in the text, words in bold underlined orange are explained in the glossary at the end. Without this safety net, the first fifty pages would bloat horribly as I tried to explain all those bits. Use the glossary when you must, but avoid it when you can. If it isn't in the glossary, look it up on the web, and then you will realise why I left all of those long technical terms in the text.

A preamble about a scramble.

In 2017, I was part-way through writing a book, tentatively called *Not Your Usual Rocks*, when I went to Sri Lanka to look at the site of the 2004 Boxing Day tsunami. Stopping at Polonnaruwa, I saw several massive statues of the Buddha, all carved from a local rock, usually called "banded granite". While I was trained as a botanist, I once studied a bit of geology, and so I smelt a rat. To my aging botanist's eye, the rock was gneiss, one of the metamorphic rocks.

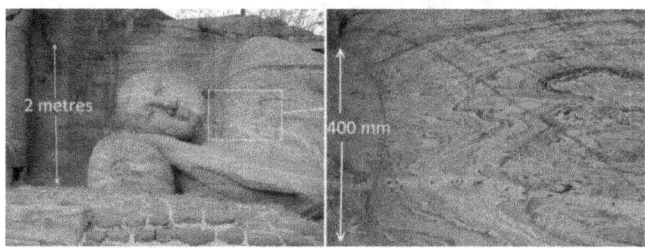

Reclining Buddha carved from local rock, Polonnaruwa, Sri Lanka.

Some 60 km away by winding road, there is a large rock called Sigiriya, a name derived from the Sanskrit *Sīnhāgiri*, meaning Lion Rock, rather like *Sinhapura*, the Sanskrit for Lion City, which is now Singapore.

About 1500 years ago, when blue body paint was the height of fashion among my ancestors, an advanced Hindu culture established a fortress on top of Sigiriya. As tourists do, we climbed to the top, and I saw the stone, close up.

Sigiriya Rock, Sri Lanka. What sort of rock is it?

Sigiriya is built on a lump of gneiss, but many web sites say the stone is granite, and a few assert that it is a volcanic plug—which would make it basalt! I had looked this up before going there, and found that the many contributors to Wikipedia dodge the issue, simply calling Sigiriya "a massive column of rock nearly 200 metres high". I knew in advance that I needed a closer look.

Clearly, 'granite', technically an igneous rock, forged in the planet's fiery depths, means different things to different people, especially those with little in the way of geological training. It really isn't rock(et) science, though.

On the left, granite from Norway, right, two Australian basalt cobbles about 19 cm long.

To a poet, any hard rock is granite, while a stone mason calls any rock with visible crystals granite, but geologists divide those big-crystal rocks up into granite, granodiorite, diorite, gabbro and more. The poet's granite and the mason's granite may not be granite at all, and a friend (not a geologist) says he once read that most Australian "granites" are really a relation, granitic porphyry.

On top of Sigiriya rock, I recalled writing a story about the Rosetta Stone while wearing my science journalist hat, some fifteen years before my climb. The tale began with an Australian quiz show causing a bit of fuss and bother after somebody found an apparent error, made a year earlier.

A contestant had said the Rosetta Stone was made of obsidian, but the quizmaster ruled that it was in fact basalt. The producers later conceded that it was "really granite", and that the contestant should have another chance to win the big money. The three named rocks are quite different, so what was going on?

The Rosetta Stone was carved in 196 BCE, and it carries three inscriptions, saying the same thing in Greek, in Egyptian demotic script, and in hieroglyphics. The content is fairly boring, a list of taxes repealed by Ptolemy V, but the use of three languages made the stone very exciting when it was found in 1799 by French forces fighting in the Napoleonic Wars in Egypt. When the French lost a major battle, the stone became a prize of war, handed over to the victors, and placed on display in the British Museum in 1802, where it remains.

The Rosetta Stone, the key to decoding hieroglyphics, was described by its original French finders as *'une pierre de granite noir'* ('a stone of black granite'), but these were not geologists speaking. When Egyptologists called it "black granite", they just meant a dark, fine-grained granodiorite from Aswan.

A geologist's granite has large and very obvious crystals of quartz, orthoclase and other minerals like mica. Forming at a great depth, it cools very slowly, leaving enough time for large crystals to develop. Granite is typically 40% quartz (silicon dioxide), and that means it is generally pale in appearance.

Basalt, on the other hand, is black to medium grey, and while it may contain a few larger bits of the minerals olivine and plagioclase, it is typically aphanitic, a term geologists use to indicate igneous rocks in which the grain size is small (less than 0.5 mm), so any grains or crystals cannot be seen with the naked eye.

That leaves us no further forward: why was a piece of granodiorite labelled basalt? Granodiorite is similar to granite, because it has quartz and plagioclase (with no orthoclase), but it also contains biotite and hornblende. It is typically darker than granite, but still a long way from basalt.

British scientists always called the Rosetta Stone basalt, and probably nobody gave the rock much thought, because it was the text cut into the stone that mattered, not the rock itself. When the stone was cleaned in 1998, it was found to be covered with black wax, printer's ink, used to obtain contact-prints of the inscriptions, finger grease and dirt, with white paint in the incised lettering to make it stand out. Calling the blackened stone basalt made some sort of sense.

Andrew Middleton and Dietrich Klemm saw that the cleaned stone was not basalt at all, and they published their findings in 2003. In chemical terms, they said, the stone is more like tonalite (that lost me as well, so don't worry!). If you want to be precise, the Rosetta Stone is made of granodiorite that has probably been exposed to some extra heating. It isn't basalt, but it should neither be taken for granite, nor mistaken for granite, either.

Now you know where the title came from for this book written for people who enjoy curiously searching around simple marvels. If that describes you, then you are in good company, because not long before he died, Isaac Newton (1642–1727) may (or may not) have said:

> I don't know what I may seem to the world, but as to myself, I seem to have been only like a boy playing on the sea-shore and diverting myself in now and then finding a smoother pebble or a prettier shell than ordinary, whilst the great ocean of truth lay all undiscovered before me.

People usually trust this quote, but Newton's interest (or lack of it) in rocks matters little: plenty of other people enjoy wandering along, looking at the rocks and bothering them, especially the ones laymen might call unusual, and that was implied in my original title, *Not Your Usual Rocks*.

So what makes a rock 'unusual', and why are rocks as they are? The 'why' question is the key one, because when you look closer, there are no unusual rocks, just interesting rocks, along with some uncommon rocks, and they all tell a consistent story about the past.

So if you like, this book is about interesting and uncommon rocks and the science behind them. Here's a quick look at some of the most interesting rocks to poke and bother.

Seven curious rock types

1. Rocks that push in between other rocks (and sometimes disappear leaving a gap). These are igneous dykes, basaltic rocks that have pushed into a joint (a crevice) as melted rock. See chapter 1.

A dyke at 1080 Beach, NSW, and a weathered-out dyke between Gerringong and Kiama, NSW.

T

2. Then related to the dykes, there are the hot rocks that flow like honey. My fondness for rocks extends to wandering out 6 kilometres over fresh (but solid) lava to see shiny, glowing melted rock. This sort of thing is also covered in chapter 1.

The author near a lava flow, Kilauea, Hawaii, in 2005. It was hot and running downhill!

T

3. Rocks with curious shapes or colours, usually as the result of something that geologists call weathering: these are discussed in chapter 3.

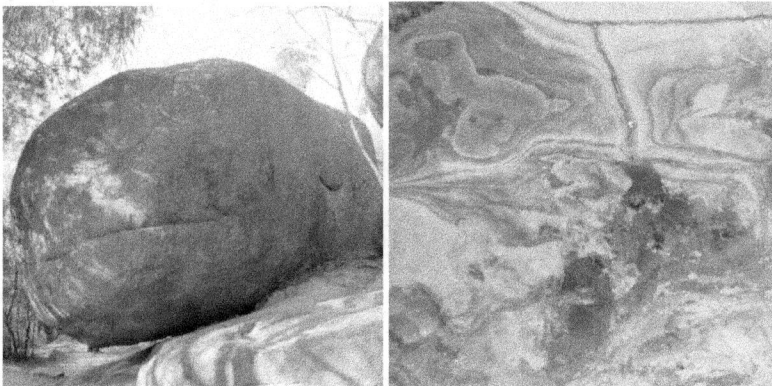

Granite boulder, Freycinet Peninsula, Tasmania, Liesegang weathering, Swansea Heads, NSW.

4. Rocks that give way to natural effects, rather like the granite above. This sort of thing is also weathering, and it is also explored in chapter 3.

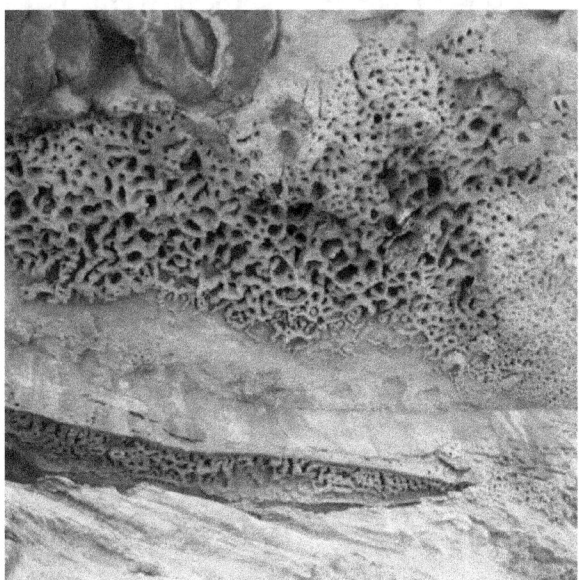

This picture shows sandstone with honeycomb weathering higher up, and cross-bedding below, Sydney area.

5. Rocks that bend, or break and slide, something discussed in chapter 5.

Folded metamorphic rock at Mystery Bay, NSW and folded, faulted rock at Aragunnu Beach, NSW.

6. Rocks that do not lie flat and level, like this case of cross bedding in Hickson Road in the city of Sydney. Cross bedding is explained in chapter 5.

This is Triassic sandstone in the middle of Sydney.

7. Rocks that float. This is what really triggered me to write the first version of this book, so I could draw attention to, and explain, the many novel rocks that people miss noticing. The seeds were sown in late 2013, when pumice washed up on the east coast beaches of Australia, delighting curious minds.

Small pumice fragments floating in water.

That arrival, and in due course, this book, were both triggered indirectly by a series of distant and unseen events involving a seamount in the Kermadec Islands, in the middle of July, 2012. Except for a small station on Raoul Island, some 250 km NNE of the seamount, this particular bit of New Zealand's territory is uninhabited, and nobody knew there had been a minor volcanic eruption there. It was a bit like that alleged winner of a British contest in 1929 to find the world's dullest headline: "Small Earthquake in Chile: Not Many Killed".

The relevance (or not) of distance in seismic events depends on where you are standing, because an earthquake can have a long reach. A Chilean earthquake in 1960 killed 5000 Chileans, and through a tsunami, 61 more people in Hilo (Hawaii), 140 in Japan and another 32 in the Philippines. The Kermadec tremors in 2012 were truly small and harmless, except to fish. It wasn't a small eruption, but it caused no tsunamis.

A previously unknown submarine volcano on L'Havre Seamount pumped out pumice, a bubble-filled rock that floats, making a raft that covered an area of 19,000 to 26,000 km^2, a piece of temporary real estate larger than Israel, lying east of the International Date Line at 30° South.

Nobody noticed the eruption while it was happening. Later, a search through seismological records identified the raft's birth date as around July 18 to 19, when there were a number of tremors in the 3.0–4.8 range. Satellite imagery showed an ash plume that went on for several days after July 18.

So why did nobody notice these quakes? The US Geological Service says there are about 13,000 earthquakes each year in the 4 to 4.9 range, and 130,000 in the 3 to 3.9 range, and who feels those? Unless the ash plume was visible, the Raoul station would remain unaware of the eruption.

During the 2010 eruption of Eyjafjallajökull on Iceland, the only airport in Europe that was never closed by the volcano's ash was Iceland's Reykjavik airport, because the ash cloud never blew in that direction. In Viking times, the citizens of Reykjavik might have remained unaware of the volcano erupting until travellers came west to tell them of it. At times, volcanoes can be surprisingly hard to notice.

Pumice is formed when gas-rich molten rock cools too quickly for the gas to bubble out and escape. Under lower pressure as magma bursts forth into the ocean, the gas expands to form vesicles (bubbles), giving a rocky froth. That sort of rock floats, and the raft that formed was up to 60 cm above the surface.

Just as most of an iceberg is under water, so is much of each lump of pumice. The actual above/below ratio can be worked out from the specific gravity (or density, if you like) of the pumice. The specific gravity (or s.g.) of pumice is around 0.64, so each piece, and the whole raft, is 64% under water. Let's call it two thirds.

In places where the pumice was 60 cm out of the water, it would have projected below the water another 1.2 metres. That said, the raft would be thinner around the edges, but we can do some rough sums.

If we assume a low value for the average thickness of the raft, just 60 cm, and take the smaller estimate of area, the volume was (19,000 x 1000 x 1000 x 0.6) cubic metres. Multiplying that by the s.g., the mass of rock would have been something like 7.3 billion tonnes. The world's annual coal production is about 7.5 billion tonnes, so that's quite a lot of rock, and we never even heard it arrive!

In the summer of 2013–14, the east coast of Australia saw large amounts of pumice stranded on beaches, and six years later, as I write this, there are still small pieces of pumice on our beaches, usually well above the tide mark. As a biologist by original training, I looked at these and found all sorts of life, including goose-neck barnacles, bryozoans and tube-worms (probably *Galeolaria*) on them, and I started collecting some of the larger pieces to photograph.

Pumice and its coating of marine life. For scale, the tube diameter is about 2 mm.

I am of a jovial grandfatherly demeanour, so when assorted grandchildren, grand-nieces and grand-nephews saw me gathering up stray rocks they wanted to know what I was doing. So did a number of other children, though in these days of stranger-danger awareness, they brought guardian adults with them, and I got the adults involved as well. As an old educator, that was fine by me.

That was how I noticed that the idea of rocks that float was an attention-grabber, and every person I talked to just had to throw a piece of pumice into the water. That got me thinking about some of the other rocks which "break the rules", meaning they fail to behave as accepted wisdom predicts.

Clearly, what rocks do *always* lies within the "rules". If we think the rules are broken, *it is our understanding of those rules that is at fault*. Still, to the lay person, a lot of rocks do unexpected things, and the seeds of this book were sown when the pumice came ashore. Still, the seeds lay dormant for a while, as I collected examples and photographs. *

I started a new notebook on October 1, 2015, and started writing on January 1, 2016. The places mentioned here reflect my travel choices, more than anything else, but the centre of my interest always comes back to the Australian continent. I actually stopped work on this book in March 2016 to do a couple of other books first, and then to race around Scandinavia, Sri Lanka, Iberia and Morocco, spotting even more rocks. Other books jumped the queue, and more travels…

The first and main interruption came when I was side-tracked by the National Library of Australia, who offered me a commission to write a book for younger readers on related matters. *Australian Backyard Earth Scientist* is now published, but there was a lot of more detailed stuff I wanted to share as well. I realised there was more to my story than rocks, so I consulted my friends. Losang Zopa, Anne Smith, Anne Graham and Peter Chubb repeatedly and iteratively played with themes, and so my new title emerged. They are all to blame…

The careful reader may see patches of similar text in the two books, because I occasionally pillaged an early draft of this work when I was writing *ABES*. That said, both books are suited to agile minds of all ages, but while pre-teens may struggle with the science in this book occasionally, it shouldn't be *that* hard.

---

* As I entered the final run on this book in late 2019, I learned that there was another mass of pumice heading towards our shores. In early 2020, I am still waiting…

Background to earth science.

South Head, Sydney, Triassic Hawkesbury sandstone.

What is earth science about?

Earth science is the only way to understand how the world works, so we can protect ourselves from things going wrong.

Once upon a time, all our knowledge came from accepted wisdom, which was bad news for real wisdom. Accepted wisdom means ancient ideas conceived by unscientific minds: there was no science back then, so the ideas just had to be unscientific. People were smart enough to want to work out what made things tick, but they relied on their own experience across short life spans.

According to the accepted wisdom of humanity's main religions, the world never changes. Nothing on our planet alters, we are told, over and over again. In *Psalm* 104:5, we read:

> He established the earth upon its foundations,
> So that it will not totter forever and ever.

Again, in *Ecclesiastes* 1:4, the King James Bible tells us:

> One generation passeth away, and another generation cometh: but the earth abideth for ever.

Even poets (people with strange notions about granite) get into the same same-old act, and Alfred, Lord Tennyson, in *The Brook*, has a stream speak like this:

> For men may come and men may go,
> But I go on for ever.

Now back to the Biblical bit, around 1640, Bishop James Ussher, using traditional ages and dates found in the Old Testament, worked out that the world began in October, 4004 BCE, .

We will return to this estimate in chapter 14, but in calculating that date, Ussher was merely repeating a view, an accepted wisdom which was widely held before his time, though he did so with rather greater precision. William Shakespeare, who died in 1616, reflected the same view when he wrote in *As You Like It*, the line: "The poor world is almost six thousand years old…"

The rocks tell us a rather different story, if and when we look at them the right way. This book is about reading the rocks' story and making sense of it. Exploring the scenery, and understanding the whys and hows of the scenery is a far more interesting notion than the idea that everything stays the same.

In my spare time, I am a volunteer in a sanctuary on a headland in Sydney, on the east coast of Australia. I help to control the weeds, heal the land, and plant new habitat for bandicoots. Any soil derived from Sydney's characteristic Hawkesbury sandstone is a poor and challenging place for life, but this is doubly so on a headland facing the Pacific Ocean. It features an impoverished sandy soil, where salt-spray-laden winds blow up and over it in every storm.

All over the continent, much of Australia's soil is old and deficient, bad news for plants seeking a home. When novelist Anthony Trollope visited Australia in the early 1870s, he looked over the sandhills around Perth and wrote:

> An ingenious but sarcastic Yankee, when asked what he thought of Western Australia, declared that it was the best country he had ever seen to run through an hour-glass. He meant to insinuate that the parts of the colony which he had visited were somewhat sandy.
> —Anthony Trollope, *Australia and New Zealand*, 557 (1873).

Trollope, by the way, was given to recycling jokes he found in old books, and it seems the original maker of this joke was probably no Yankee at all:

> Fremantle, of which it was wittily said by the quartermaster of one of His Majesty's ships who visited the place, "You might run it through an hourglass in a day," is but a collection of low white houses scattered over the scarce whiter sand.
> —John Lort Stokes, *Discoveries in Australia*, Volume 1, chapter 1.3. (1846)

You mightn't expect it, but poor and unfriendly places like the white sandhills of Western Australia and the thin, washed-out sand on Sydney's headlands offer challenges that force plants to find micro-niches they can cling to.

Poor soil like this generates an amazing biodiversity, because each niche varies in some subtle way that marginally favours one plant over all others because that one has broader leaves, deeper roots or thicker cuticle—or the niche may subtly favour some animal with larger lungs, a different colour or longer legs.

These specialisations place that animal or plant at a disadvantage if it moves, compared with its neighbour, because the second plant or animal has its own special attributes. Each thrives in its own favoured and favouring spot, so similar but different species emerge, side by side, and the biodiversity index goes up.

We are still finding new species up on my play space, Sydney's North Head. These aren't species new to science, just new to our records, but it is a fascinating and rewarding place to work, and I learned, soon after starting there, that the headland was covered with deposits of white sand, blown there in the last ice age.

I also learned that one academic believes that a tsunami dumped all the sand on top of the headland, though he seems not to have explained how the sand was dropped with precision accuracy, and then persuaded to remain in place as the waters cascaded off the headland. Trust me, the sand is aeolian: it blew in. All one needs is some common-sense logic.

The sanctuary has gathered in an remarkable set of talents and minds, making it an interesting place to work. What I bring to the party is a training in biology and a lifetime of work as a communicator of science. I also bring curiosity and an understanding of rocks and soil.

In my day job, I write about science because I have a curious mind. I am a cross between Kipling's mongoose Rikki-Tikki-Tavi, who always had to run and find out, and his Elephant's Child who had "*'satiable curtiosity'*". Kipling used this verse to end his fable about why elephants have trunks:

> I Keep six honest serving-men:
> (They taught me all I knew)
> Their names are What and Where and When
> And How and Why and Who.
> —Rudyard Kipling, 'How the Elephant Got Its Trunk', *Just So Stories*, 1902.

For me, it isn't enough to know that animals eat plants (or that they eat animals which eat plants), or that plants grow in soil. I want to know where the soil came from, how it formed, why it is there and more. That means I care about rocks, and I know there is more to rocks than just being a solid foundation to build on.

The day I set out to write my first rough outline of this book, I mentioned my plans to a colleague. Our chat expanded into a few of the areas I hoped to look at, and as often happens, I began talking about the sandy soil. Jenny said, idly, "And of course, all this sand was once sandstone."

I agreed, but added "…and not only that: all this sandstone was once sand, and before that, I'm told it was a different sort of stone, somewhere near Broken Hill". She thought about it and supposed it must be so, now I mentioned it. She simply had found no reason, before our discussion, to ask where the rocks came from. Unless you think about it, the rocks are just there, and usual…

We went off on different tasks, but as I set to work blocking further erosion in a developing gully on a track, I thought more about this discussion, because simple physical work allows one time to think, and I realised we don't think enough about the rocks that shape our scenery. That night, riding to Sydney on the ferry for a concert, I started to jot down some headings about explaining geology.

As we approached Circular Quay, it struck me that when he first went on board HMS *Beagle*, Charles Darwin was more of a geologist than anything else. All his evolution stuff came later—but without the rocks, there would have been no life to evolve. Rocks, and the curious rocks, the not-your-usual rocks were indeed a worthy topic for further enquiry.

As we left the concert hall, several hours later, I quoted Paul Ehrlich (1854–1915), a chemist who once said to *his* wife after a symphony concert:

> Those really were two unforgettable hours. It's been a long time since I've been able to concentrate so well on my problems with arsenophenylglycine. We'll have to make a small substitution the first thing tomorrow.

I refused flatly to explain why that seemed apposite, but like me, she's a science-trained person, so she worked it out.

A major strand in this book deals with how scientists work things out. The science will be gentle, and explained, but be aware that science lurks under every rock. Mind you, if you look carefully, some of the science will also be sitting on the rocks, sunning itself. Science is like that.

In the 2020s, climate is the part of earth science we simply *must* understand, but earth science began with the rocks and geology, and so shall we.

What geologists agree on.

It is the customary fate of new truths to begin as heresies and to end as superstitions.
—Thomas Henry Huxley, (1825–1895), *The coming of age of The Origin of Species* in *Science and Culture* xii.

Almost every city in the world is shaped and defined by its geology, in all sorts of unexpected ways. Some of the effects are easy to see, like the large volcanic remnants which feature in central Edinburgh, the soft clay of London which allows tunnels to be drilled for underground railway lines, or the chalk of Paris which allows the Métro lines to be put in place.

Around my home in Sydney, sitting on top of a 200-metre-deep bed of sandstone, the area is shaped by the geology. The hard sandstone and drowned river valleys of Sydney, produced by a rise in sea levels at the end of the last ice age, guarantee a sprawling city with many bridges.

A sandstone cliff near Fairy Bower, Sydney with joints (planes of weakness that we will visit later).

The fern-leaf pattern of Sydney harbour comes from the mostly north-south and east-west joints in the sandstone. When the sea level was much lower, during past ice ages, Australia wasn't frozen, and water seeped in and ran along the joints, carving gullies and valleys that were later filled in ("drowned") by the rising seas.

Some cities like Budapest were shaped by the rivers they lay on, rivers that were in turn shaped by geology. Then again, the Maltese people survived prolonged bombing in World War II, and their island provided an important strategic base for the Allies against the Germans, because the population could shelter in limestone caves. Liverpool in England only became an important port after silting of the River Dee ruined Chester as an ocean port.

Every cliff and rock face, even most pebbles, will carry a tale on and in it. The rules of thumb we use to explain rocks are the rules of scientific geology, but science is variable in its quality and in its actions.

Beach pebbles, Gerringong New South Wales.

The more general a scientific rule-of-thumb is, the more scientists agree about it. But no rule is sacred: if somebody finds just one contradiction, people need to think, and toss out a part, or even all, of the old rule.

Sometimes, two scientific models can rest happily, side by side. The 333-year-old (in 2020) physics of Isaac Newton is good enough to send rockets to the moon or the outer planets, but to explain the universe, we need the more modern ideas of Albert Einstein.

These days, adjustments to science seem to happen less often, but it's always possible that one of the foundations, one of the basic assumptions of geology needs to be altered or discarded. How much story can you see in the picture below? I see lots of past history, but then I was trained by clever observers and I've been looking curiously at rocks for six decades—and I know what geologists agree about. Let's go there.

Hawkesbury sandstone cliff, North Head, Sydney.

Agreement 1: We can see where the world has changed.

The world does *not* abide forever, even if, in our short life spans, the world really *is* fairly permanent, because the occasional volcanic eruption or a massive earthquake can make a difference to what we see. Floods can move huge amounts of sediment as mud, and the occasional collapse of a delicate eroded arch or cliff may be seen, and rivers may alter their courses in a flood, but that's about it for changes in any single human lifetime.

Most, but not all, geological changes are *very* slow and hard to spot. An example of a rapid change was in November 2016, when an earthquake near Kaikoura, north of Christchurch in New Zealand, suddenly lifted a long strip of seabed above the high tide mark.

By 2020, all you could see was a strip of rocks, whitened by the dead marine life on them.

On a larger time scale, continents shift around, pushing and jostling. As they do, they shove the Swiss Alps, the Himalayas, the Andes and all the other mountainous parts of the world up into the sky. At the same time, weathering destroys rocks, and erosion carries the remnants down to lower reaches. Without new rocks being pushed up, there would be no mountains left.

Sometimes, land sinks deep beneath the sea, and new material gets washed in and laid down on top, compressing the sediment to rock. Later, some of those new underwater rocks may be shoved back up into the air. All geological-scale events are slow, and we know this because we can work out the dates of rocks.

The secret lies in noting the slow processes and imagining how massive their effects might be over a large enough period of time. Once they did that, people needed to reconsider how old our planet was, and how old the rocks might be.

Agreement 2: We can tell the ages of rocks.

Our various dating methods often give slightly different results, because geochronology, the dating of rocks involves inexact measures. Sometimes the assigned date relies on inference or assumptions, like the cases where we find fossils that come from a species that only lasted a short while. We say that when similar layers in two cliffs carry the same index fossil, they are the same age.

There can be traps. We may have misidentified one of the fossils, or we may be wrong about how long that species survived. The good news is that we keep finding new methods and new data, so over time, the picture becomes clearer. The better news is that the adjustments are generally small, because the different methods all give a consistent picture.

At times, we may use the half-lives of radioactive minerals, or other measures: once again all of the methods give *consistent* results. The *order* of formation given by different dating methods is the same, and over time, we have got much better at putting precise year-counts, on things. Our planet formed around 4.6 billion years ago, though the sandstone I walk on most days is Triassic in age, and roughly 200 million years old, but the sandstone won't last forever.

Agreement 3: We can see that rocks break down.

Rocks don't abide forever: they are less than everlasting. Their minerals break down, under the combined effects of water and air, and the rocks come apart under the mechanical effects caused by heat, cold, grinding rocks carried by rivers and *glaciers*, and even from sand-blasting in deserts when strong winds blow.

Then there are the biological influences on the rocks. Burrowing animals from ants to echidnas drag grains of partly-weathered material to the surface where they are more exposed to sun, water and rain.

A digging animal, an echidna, *Tachyglossus aculeatus*, 10 km from the centre of Sydney, and an animal pad in arid country near Oodnadatta provides easier walking for later animals.

On the surface, large grazing animals make pads, tracks along the sides of hills, pushing sediment down and providing a path for water to run off downhill when it rains, carrying surface sediment away. On a personal note, I don't exclude my own walking on sandstone from this.

Tree roots grow into cracks in rocks and expand, splitting the rocks, while at the other end of the scale, some mosses can drill neat holes in quartz, one of the toughest of minerals. In bulk, the Earth abideth well, and nothing is ever lost—it just bobs up in a new hat.

Agreement 4: Geology involves reusing old material.

None of the material coming from rocks is ever wasted. Calcium from basalt will eventually be dissolved and carried to the sea, where corals, snails or other marine life will extract it and use it to make skeletons, shells or something else. Later, these dead animals may fall down and over time, become limestone.

North of the Hawkesbury River's mouth, NSW: muddy water is carried northward by currents, before it settles.

On land, the calcium may pause to form cell walls in plants, the shells of birds or the bones of vertebrate animals. Sometimes, the shells just dissolve once more, but over time, all the calcium is washed down to the sea. Sand becomes sandstone, mud becomes shale, shells make limestone, and so on.

If rocks get buried deeply enough, they may be changed by heat and pressure, so sandstone becomes quartzite, limestone becomes marble, and shale becomes slate. If the rock is heated enough, it may end up as molten lava that spews out onto the surface of the earth again. Rocks aren't all "just rocks", because the world's rocks have many different origins.

Agreement 5: Rocks come in three main types.

If you blinked and missed it, I referred in the last section to three different sorts of rock: the ones made from sediments, the ones shaped by heat and pressure, and the ones that were melted before they became rocks.

Sedimentary rock forms when sediment particles (bits and pieces of almost any sort) fall to the bottom of a lake or sea (or sometimes the bottom of a vast sand dune). Later, the particles are buried and a complex mix of pressure, water washing through and maybe some heat, turns it into a rock. Sedimentary rocks are the ones that sometimes contain fossils, though a few deformed fossils may also be seen in slate or marble.

Igneous rock is any sort of material that was once a hot liquid, because at high temperatures, all rocks will melt. Granite forms far below the earth's surface, and only appears when a whole load of surface rock erodes and weathers away. Granite cools slowly enough for big mineral crystals to form. Basalt, on the other hand, is oozy stuff that flows and spreads. Some basalt comes out of volcanoes as lava, some oozes out and flows across the land, making a flat sheet that cools fast, so the crystals are very tiny.

Some of the magma pushes through the cracks that geologists call joints, but joints will come later, and so will the dykes, formed by the cooled magma.

Metamorphic rock (the name comes from Greek, and means *changed shape*) forms when another kind of rock is subjected to heat and/or pressure. Large metamorphic areas are usually formed by extreme heat and pressure, and this is called regional metamorphism. When melted lava flows over other rocks, or when a volcano pushes through other rocks, the heat may travel a few metres or tens of metres, causing contact metamorphism.

In short, there are very few inexplicables when it comes to looking at the rocks, because in the end, the old rocks are broken down and wiped out. Still, you can tell when an inexplicable shows up, because the scientists start formicating.

No, read that again, carefully. Scientists know that *formication* means running around like ants in a panic. Spelling is important in science, and many other things: consider the famous and possibly apocryphal one-letter typographical error when a newspaper was supposed to report that Queen Victoria "...*passed* over the bridge to the accompaniment of thunderous applause".

We all agree that was funny, and we can even explain why, because humour, like geology, is explicable, if you know the right things.

Agreement 6: All geological effects can be explained.

Basically, all the things we see in the shape of the world can be explained by the forces we see operating today. Geologists call this the principle of uniformitarianism, which just says the natural laws and processes we see shaping the earth today are the same ones that shaped the past.

In other words, we don't work on the idea that there used to be wizards and witches who moved the rocks around, that there were fire-breathing dragons which made the lava melt. We don't need to assume the existence of pixies driving Stealth Bulldozers or poltergeists with geological habits. Yesterday's causes of change are today's causes of change.

The rocks, even the not-your-usual rocks like these, keep on following the rules.

Continents move, floating on the surface of the planet; earthquakes happen; rocks form, then they weather and are carved by erosion; rocks get pushed up; others are pushed down and buried, and so on. On a smaller scale, sediments get washed away by water, blown around by winds, or forced along by glaciers. They always have done, and for some considerable time, they always will.

When weight is applied to the existing surface, in the form of glaciers or any other way, the earth's crust behaves like a small raft that an elephant has boarded. Like the raft, the rocks sink. On the other hand, when glaciers melt, the earth springs back up again, and this is currently happening in Scandinavia which was relieved of a lot of glacial weight, about 10,000 years ago. Geology never rushes…

Agreement 7: There are standard rules of geology.

Sometimes, what you see may seem contrary to the rules below, but if you think that, it usually means you simply haven't thought hard enough. The apparent contradictions emerge only because you are unaware of the other rules that apply in a particular place. With enough thinking, you can explain what you see.

Rocks are usually laid down in flat layers.

It is a fairly safe rule that sedimentary rocks form flat, parallel beds, because the sediments are washed or blown into some sort of basin, and the first material fills in the gaps and crevices, leaving a flat surface. The effects of currents (or winds) and gravity keep the top fairly flat after that.

Even when one layer is humped up, the layers that follow even the surface out.

Against that, some beds can be laid down on a slope. This effect is called cross bedding, and we will look at it in more detail in chapter 5. Cross bedding can be distinguished from beds that have been tilted later by looking for the horizontal beds above and below.

Cross bedding in Hawkesbury sandstone, Old Man's Hat, North Head, Sydney, Australia.

There can be traps for the unwary when it comes to igneous rocks. If the rock arrives as lava, streaming down the flank of a volcano, some of the lava cools and becomes solid, leaving a sloping skin of rock.

Eroded remnants of an old volcano near Cape Palliser, North Island, New Zealand.

Younger rocks usually lie on top of older ones.

Rocks are always laid down that way, but there are a couple of notable exceptions. Magma sometimes pushes up through sedimentary (or other) rocks to form a dyke. If the dyke reaches the surface, it flows out over the landscape (when it is called a flow). A flow is always younger than the rocks it lies on top of, and older than any rocks which are found above it.

Sometimes the magma pushes between two layers of rock, forming a sill, but the igneous rock remains younger than the rocks that lie on top of it. How do we know what happened? We look for contact metamorphism, above and below the layer.

The other exception to youngest-on-top comes when rocks bend, and fold, and sometimes (very rarely), overfold, so that the usual age order is reversed locally. In less extreme cases, horizontal beds may just be tilted up and eroded away, leaving sloping beds behind. If the land sinks at this point, new sediments wash in to start a new age of rock building. We will come back to this shortly.

In an area where there are active volcanoes, lava may pour out and flow across the countryside, laying fairly flat layers—except, as mentioned above, on the flanks of the volcanoes, where sloping beds form.

The Columbia River forms the border between Washington and Oregon in the USA. It flows through a valley carved through a massive series of basalt flows.

⚒

There can be gaps in the geological record in any place.

On my home territory, near Sydney on Australia's east coast, the rocks are Triassic in age. If you drill straight down you will come eventually to Permian rocks, the coal measures that are exposed around the margins of what we call the Sydney Basin. Because the beds form a basin, you find coal near the surface at Newcastle, Wollongong, Lithgow and other places. Coal also used to be mined on the very shores of Sydney harbour, but the miners had to drill quite a long way down, all the way to the Permian rocks.

In theory, if we keep drilling, we should next find Carboniferous rocks, but in the Budawang Ranges, my favourite wandering place, these layers are missing, and instead we encounter tilted Devonian metamorphic rocks. It looks as though we are missing 100 million years (or more) of geological history.

Any rock-hound will tell you this gap is an unconformity, and they will hazard a guess that the Devonian rocks may have been deeply buried and covered with Carboniferous rocks, but that the earth and its rocks moved hugely, and any Carboniferous rock was eroded, leaving ribs of tough tilted Devonian stone across the land in the early Permian era. We can't be sure there were ever any Carboniferous rocks, but it is likely they came and went, leaving no trace.

Later, the land all sank deep into a sea in some sort of cataclysm. In the Budawang ranges, west of Nowra, south of Sydney, the lowest layer of the Permian rocks is a conglomerate containing very large boulders, telling us that the first deposits in that part of the basin were laid down in a huge flood.

At Myrtle Beach, on the south coast of NSW, this conglomerate layer is missing, suggesting that the oldest Permian sediments there were laid down at a different time. It may have been a few years, more probably it was a few millennia—or even quite a few millennia. The estimated age gap either side of the unconformity here is ~200 million years, with Ordovician rocks below and Permian above.

Myrtle Beach, south coast of NSW. The sloping beds below are pointing to 1 o'clock, and the hand (top left) spans a gap of about 200 million years in the geological record.

There is also a simpler sort of time gap, much harder to identify, called a disconformity. This happens when sediment stops being delivered for a while, but we can largely ignore disconformities for the moment. We now have the basic background to understand a bit of slightly more detailed geological history.

The laws and principles of geology.

Nicolaus Steno started it. Here is a modern version that conveys his thinking in language we use today.

* Steno's Law of Superposition says that in a sequence of strata, any stratum is younger than any strata on which it rests, and older than any strata above it.

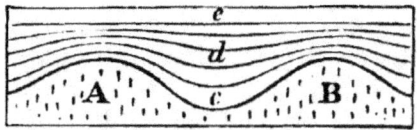

An illustration from Charles Lyell's *The Student's Elements of Geology* (1871), page 17, showing how irregularities in an underlying surface are filled in, slightly contradicting Steno.

* **Steno's Law of Original Horizontality** says strata are deposited horizontally and then deformed to various attitudes later. That is, undisturbed true bedding planes are nearly horizontal, though we need to note here that cross-bedding is possible where sandhills or sandbanks are being formed.

* **Steno's Principle of Lateral Continuity:** strata initially extend sideways in all directions. That is, every outcrop in which the edges of strata are exposed demands an explanation, and similar strata on two sides of a valley represent erosion of the rock between.

* **Steno's Principle of Cross-cutting Relationships**: anything that cuts across layers post-dates them. This applies particularly to igneous intrusions such as dykes. Aside from Steno's principles, geologists accept the following notions:

(1) An intruding rock is younger than the rock it intrudes into;

(2) A fault is younger than the rock which is faulted;

(3) William Smith's principle of fossil succession; and

(4) Any pieces of 'foreign' rock included within a rock must be older than the rock they are found in.

Xenoliths (foreign rocks) in sandstone, Kangaroo Island, South Australia.

We will come to Smith's fossil succession soon, but geology was only possible because of James Hutton. He had made enough money from his ammonium chloride factory to be able to retire from work and study geology.

Hutton was an old friend of Joseph Black, the first scientist to distinguish heat from temperature, and also a friend of James Watt (of steam engine fame), so it is no surprise that Hutton assumed that all earth activity was due to what he called the earth's 'heat engine'. Most importantly, he said "…The past history of our globe must be explained by what can be seen to be happening now".

He emphasised the igneous origins of rocks (unsurprisingly, given that he came from Edinburgh, where igneous rocks are common). The French Revolution was happening, so the public of Britain were unenthusiastic about Hutton's revolutionary notions. They were unprepared for his ideas, and unwilling to accept them, but soon, still in the early 1800s, they were ready.

John Playfair was trained in mathematics at a time when geology had not been invented, so he was necessarily largely self-taught. Like James Hutton, Playfair was exposed to the stimulating geology of Edinburgh, which would have assisted him in his thinking.

He invented geomorphology, giving us 'Playfair's Law', which says rivers cut their own valleys. He also said the angle of slope of each river shows an adjustment towards a balance between the velocity and discharge of water on one hand, and the amount of material carried on the other.

Playfair made the work of Hutton more accessible with his *'Illustrations of the Huttonian Theory of the Earth'* in 1802. Playfair also explained the rock cycle of repeated weathering, erosion, deposition and solidification in simple terms:

> The series of changes which fossil bodies are destined to undergo, does not cease with their elevation above the level of the sea; it assumes, however, a new direction, and from the moment that they are raised to the surface, is constantly exerted in reducing them again under the dominion of the ocean. The solidity is now destroyed which was once acquired in the bowels of the earth…
> —John Playfair, *Illustrations of the Huttonian Theory of the Earth*, 1802, 109.

His ideas gained wider acceptance after Charles Lyell added Playfair's thoughts into his *Principles of Geology*, but we have ignored William Smith long enough. An orphan, he was set to work early as a surveyor for the new canals that were crossing the British countryside, so industrialists could haul goods around.

These canals needed digging into the ground, and even tunnels through hillsides, and this gave Smith first-hand chances to observe and classify the many rock types as they are seen in fresh unweathered exposures. Most importantly, he saw how strata were typified by fossils, and he pointed out that the same stratum could be identified over a considerable distance by the fossils it contained.

In 1816, Smith published his ideas, accompanied by a coloured geological map, and made the point that, given the law of superposition, the fossils in the strata gave us a view of the history of life on earth. Now the way was open, and Charles Lyell's *Principles of Geology* could be released in the early 1830s, just in time for Charles Darwin to take a copy with him on the voyage of HMS *Beagle*, so he would be prepared to unravel in full detail the reasons why life actually possessed a history on earth.

That is how science weaves itself into a web, but science also involves cycles.

Rock material gets recycled.

Modern geology is all about the rock cycle, with material being raised to the surface by volcanoes to be weathered, eroded, washed and deposited to maybe later become sedimentary rock, or heated and compressed to metamorphic rock, or melted to form igneous rock, which may later return to the surface again. Until subduction zones were observed in plate tectonics, where one plate slides beneath another, the burial of rocks in the rock cycle was something of a mystery, but this is now much clearer.

Long before Twitter, I wrote a set of simple scientific principles that were limited to 160 characters. Each proposition had to stand on its own, in 160 characters or less. The aim was to see if there was a simple way to define science and all its curiosities. They are still on the web as "Science SPLATS", a name I gave them before I worked out what SPLATS meant, and I never got beyond "**S**cience **P**rinciples, **L**aws, **A**ssumptions, **T**heories and **S**omething-or-other".

It doesn't matter how the name was reached, but here are some of my SPLATS that relate to rock cycles:

> The material that we call rock goes through cycles, being melted, weathered, eroded, buried and eventually heated and compressed until it melts again.
>
> Rocks are mostly made of minerals or their weathering products, which were, at one stage of their existence, crystalline, and may still be crystalline.
>
> Rocks erode and re-form in the rock cycle. The process involves chemical and mechanical weathering, erosion, transport, deposition and compaction
>
> Rocks and soil erode, water transports sediments downstream, and the sediment particle size in a stream depends on the speed of the water flow.
>
> Sedimentary rocks buried under a sufficient load of more recent sediment may be compressed and heated so that over time they form metamorphic rocks.
>
> Most minerals in rocks are present as crystals: the size of the crystals in igneous rocks shows how quickly they cooled, with slow cooling giving big crystals.

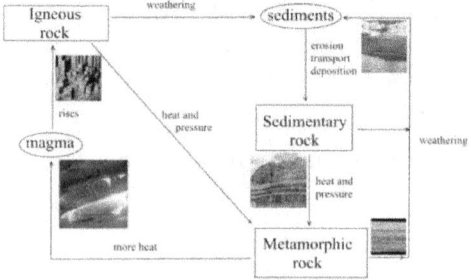

The rock cycle.

Agreement 8: The principles of science apply to rocks.

    Atoms and molecules.
    The laws of thermodynamics.
    Conservation of mass and energy.
    Equilibrium.
    The law of large numbers.
    Evolution.
    Falsifiability.
    Ockham's Razor.

These principles are what all scientists just naturally assume, but they are rarely stated explicitly for the public. There is enough information here to tell you what the idea is about, and enough terms to look the idea up, if you think you need to. There are lots of ins and outs, and working scientists spend their lives mastering them. Such people will, I hope, recognise that this is Science Lite.

Atoms and molecules.

To begin with, all matter is made up of atoms, and the properties of matter will depend on what atoms are present, as well as how they are arranged and connected. Many atoms join up in regular frameworks that we call crystals.

The laws of thermodynamics.

Putting it simply, heat flows from hot to cold, and perpetual motion is impossible. By the way, if you want to drive a politician or an arts administrator to distraction, ask him or her to explain (or even just to state) the second law of thermodynamics. Trust me: it matters!

> A good many times I have been present at gatherings of people who, by the standards of the traditional culture, are thought highly educated and who have with considerable gusto been expressing their incredulity at the illiteracy of scientists. Once or twice I have been provoked and have asked the company how many of them could describe the Second Law of Thermodynamics. The response was cold: it was also negative. Yet I was asking something which is the scientific equivalent of: Have you read a work of Shakespeare's?
> —C. P. Snow, Rede Lecture *The Two Cultures and the Scientific Revolution* (1959).

Entropy.

This is harder. We begin with the simplest available version first: entropy is a thermodynamic function that measures randomness or disorder. Or if you like, entropy is a measure of untidiness. Most of the principles of science are what scientists call *counter-intuitive*. In lay terms, they seem to go against our gut reaction: the earth as we experience it looks flat, and our intuition tells us the sun and moon circle around us once a day, but all scientists agree that the world is a globe, we orbit around the sun once a year, and the moon orbits around us once a month.

Entropy is slippery, rather than counter-intuitive, and you have to note the qualifications which limit entropy to events *inside a closed system*. Under those restrictions, entropy, or disorder, increases, which is how scientists say that over time, everything in a system gets more random, more dispersed.

There can be no exceptions to the rule that the disorder always increases, but life, at a local level, can be an anti-entropy agent, making things more ordered at that level, even as entropy is increasing on a larger scale. In simple terms, animals and plants gather up and concentrate certain elements in our bodies. In hot magma, crystals may concentrate rare elements.

Across the universe, every change leads to an overall increase in the total entropy, but the delight lies in the details, and a lot of geological science comes down to explaining how, on a local level, the process of concentration in elements or minerals is driven.

> If [*your pet theory of the universe*] is found to be contradicted by observation—well, these experimentalists do bungle things sometimes. But if your theory is found to be against the second law of thermodynamics I can give you no hope; there is nothing for it but to collapse in deepest humiliation.
> —Sir Arthur Stanley Eddington, *The Nature of the Physical World* (1928), chapter 4.

Conservation of mass and energy.

In simple terms, matter and energy can neither be created nor destroyed. There is **No Such Thing As A Free Lunch**, anywhere in science. Not ever!

Equilibrium.

As a rule, things are in balance, but that doesn't mean they are unchanging. The number of oxygen molecules in an open jar on my desk may vary slightly over time, as molecules whizz in and out, but at any practical level, the entries and exits cancel each other out. We say the molecules are in dynamic equilibrium.

The law of large numbers.

There is no such law, but it is convenient to pretend it exists. Given time, every atom of a sample of radioactive carbon-14 will break down. We cannot say when a given atom will decay, but with large numbers of atoms, we can say that half of all of the carbon-14 atoms that we start with will have decayed if we come back in 5730 years. Scientists say carbon-14 has a half-life of 5730 years. Start with a large number of ^{14}C atoms, and after 5730 years, half of them will have decayed.

Evolution.

Evolution also hangs on large numbers. You won't evolve, I won't evolve, but our species, like every other species, *does* evolve, if there are enough of them.

Falsifiability.

Every part of science is able to be falsified by evidence, and if some idea can't be tested and potentially falsified, it just isn't science. That doesn't mean science is all false, it just means *every* assumption is always considered open to testing and being found wrong If we found dinosaur fossil bones and human fossil bones in the same rock, this would mean we had to revise large parts of what we think we know about geology and biology. Scientists are always on the alert for contradictions like that, even though they don't really expect to find any. One way to become a famous scientist is by finding a red-hot contradiction to what everybody believes.

Ockham's Razor.

Then again, maybe we wouldn't need to revise anything except our viewing position. William of Ockham made it a lot more complicated, but his basic notion was that if there are two possibilities, you should take the simpler one. If we found human and dinosaur fossils in a single rock, a simpler explanation would be fraud. We would at least look for evidence of fraud first, but if there was truly no evidence of fraud, then it would be time to start a rethink.

Caveats.

I am neither an earth scientist nor a geologist, but I know how to think, where to look, and what questions to ask. My undergraduate studies were mainly in the areas of botany and zoology, so I may, from time to time, be in error. As a professional science writer, I am used to checking my facts, but even when I get the latest opinions, there is still one *gotcha* remaining.

Geological science *does* change, and I saw it happen. When I was an undergraduate, I picked up a year of formal geology training, and in 1962, one of our geology lecturers alerted us to certain sessions of ANZAAS, the Australian and New Zealand Association for the Advancement of Science. "Listen to Sam Carey," he told us. "He's quite mad: he thinks the continents are moving."

I did indeed hear Sam Carey talking about his wild ideas. He seemed to make a reasonable case, except that we all knew the idea was crazy. Yet just three years later, plate tectonics was all the go. In fairness, Sam Carey was only partly right, because his notion was based on some false assumptions, but the key thing to note is this: in 1962, believing in moving continents was madness, by 1965, it was pretty much the orthodox model.

Here, I have tried to stay with the best and safest bits of orthodoxy, but at any time, that which was orthodox can be defeated or overturned by a simple paradox. One new discovery is all it takes to change things, as T. H. Huxley said while discussing work on the spontaneous generation of life:

> But the great tragedy of science—the slaying of a beautiful hypothesis by an ugly fact—which is so constantly being enacted under the eyes of philosophers, was played almost immediately, for the benefit of Buffon and Needham.
> —T. H. Huxley, *Presidential address to the British Association* in September, 1870.

This book is about the facts—though in chapter 15, I will discuss a maverick theory about the origins of oil. I don't believe that theory, but it is both entertaining, and instructive to consider, as a way of seeing how science works.

How this book works.

Curved columnar basalt on the Snake River.

Rock melts when it gets hot, it flows downhill, it gets hard as it slows and cools off, and thick beds often form columns. When I first saw the structure above in northwest USA in 2015, I declared that I would be writing to President Obama, demanding that he send in the Corps of Engineers to blow it up, on the ground that it was contrary to the laws of physics. For the unwary, I spoke in jest.

I have since given the matter a lot of thought, because columnar jointing, with straight up-and-down "pipes" is very common in basalt flows all over the world. The Giants' Causeway in Ireland may be the most famous example, but there are plenty of others. The curving, by the way, was both known and explained by Charles Lyell, who saw similar formations in the 1820s, but I will come back to that later, in chapters 1 and 2.

Bryce Canyon, Utah, showing pinnacles called hoodoos, formed in part by the flaking action of winter ice.

Over time, rocks break down and wear away. Sometimes, this can lead to amazing forms, like the pinnacles above, from Bryce Canyon in Utah. In chapter 3, we will look at the chemical and physical changes geologists call *weathering*, and more, because weathering is responsible for the shape of the world we live in.

When rocks weather, the bits and pieces fall down as far as they can, and then they are carried by wind and water (and sometimes by ice in the form of glaciers) to some other place, and sometimes form new rocks that reveal their origins. Chapter 4 looks at the ways of moving sediments, how they become rocks and how the sandstone at Zion National Park in Utah must have formed by desert winds pushing sand around. Chapter 5 is about sedimentary rocks.

Deeper down, things get even hotter, and the pressure is even greater. Chapter 6 will skip lightly over how we know what is down there, and how rock changes in places far too dangerous for us to ever go. We will look at where we think the heat comes from, and how that ties into the age of the planet.

In chapter 7, we look at things going wrong, mainly earthquakes and volcanoes, and that brings us to igneous rocks, then we look at larger-scale movements, plate tectonics and how that affects us in chapter 8.

London's Broad Street pump was destroyed in World War II. This replica is near the John Snow pub in Soho.

There's more than rock underground, and chapter 9 is about groundwater, water tables and wells, including a very clever Iranian invention, a well that goes sideways, uphill. We will look at the pump above: the original caused a small epidemic in the 19th century, because people did not understand groundwater.

Trilobite, frontispiece from Lyell's *The Student's Elements of Geology* (1871).

Some of the things we find in rocks were once alive, and we call the remains fossils. Curiously, as you will learn in chapter 10, without water in the rocks, few fossils would form.

Chapter 11 looks at minerals and economic geology, before we turn to the many ways humans use and have used rocks for engineering, for art, and in other ways, some of them quite unexpected.

Inca wall at Machu Picchu, Peru, made without any metal tools.

When you come down to it, most of the rocks around us are unusual—and most of them can leave us puzzled, if we don't know about the ways rocks behave. We need to change that, which means readers need to understand how geologists and earth scientists think, and that comes in chapter 12.

Gravity is involved in every part of rock work, and that comes in chapter 13, before we turn to finding the ages of rocks in chapter 14, and then look at how the world may end, thanks to the political clowns and greedy idiots who are ignoring geology, physics and climate science. Chapter 15 is about why these fools should be left on a small desert island, and chapter 16 is about endings.

Chapter 17 is a short glossary of some of the technical terms, the words that appear in underlined bold orange letters throughout the book.

A planetary data sheet.

Age: 4.54 billion years.

Shape: Not a sphere, but an oblate spheroid like a pumpkin, almost spherical, but flattened at the poles.

Size: Around 6378 km in radius at the equator, mean radius 6371 km, volume 1.0832×10^{12} km^3.

Areas: Planet surface: 5.10×10^8 km^2; continents: 1.49×10^8 km^2; oceans: 3.61×10^8 km^2.

Heights and depths: mean height of continents, 875 m, mean depth of oceans 3794 m.

Gravitational pull: The acceleration due to gravity at Earth's surface is about 9.78036 metres per second per second (ms^{-2}) at the equator and 9.83028 ms^{-2} at the poles, lower as you get higher, so that at 1000 km up, it is 9.08675 ms^{-2}.

Mass: Approximately 6×10^{24} kg, mean density 5.515 (Henry Cavendish measured the mass indirectly with a torsion pendulum, the rest is calculated.)

Layers: There is an outer silicate crust, averaging about 35 km deep (varies from 5 to 70 km), that is solid; a viscous mantle from 35 to 2890 km down; a liquid outer core that is more runny than the mantle (2890 to 5150 km); and a solid inner core. (Most of this information comes from analysing earthquake waves.)

Component masses: atmosphere 0.0000051×10^{24} kg; oceans 0.0014×10^{24} kg; crust 0.026×10^{24} kg; mantle 4.043×10^{24} kg; outer core 1.835×10^{24} kg; inner core 0.09675×10^{24} kg. (All estimated by clever science.)

Above-ground layers: The planet has a number of gaseous shells around it. The troposphere is where all the weather happens, and that goes up to about 12 km, though it is higher near the equator. Above that is the stratosphere, up to 50 or 55 km, the mesosphere and other layers go out to 10,000 km. There are several shells known as the magnetosphere, and these protect us from cosmic radiation. Beyond that, it is space. (Most of our knowledge comes from early balloon flights and later rocket flights, carrying instruments.)

Water: 97% of the water is in the seas and 3% is fresh. (These can be calculated.)

Elements in the crust: oxygen 46.6%; silicon 27.7%; aluminium 8.1%; iron 5%; calcium 3.6%; sodium 2.8%; potassium 2.5%; magnesium 2.1%; rest 1.6%. (Obtained by measurement and estimation.)

Elements in the planet: iron 34.6%; oxygen 29.5%; silicon 15.2%; magnesium 12.7%; nickel 2.4%; sulfur 1.9%; titanium 0.05%. (All estimated.)

1: How the earth was made.

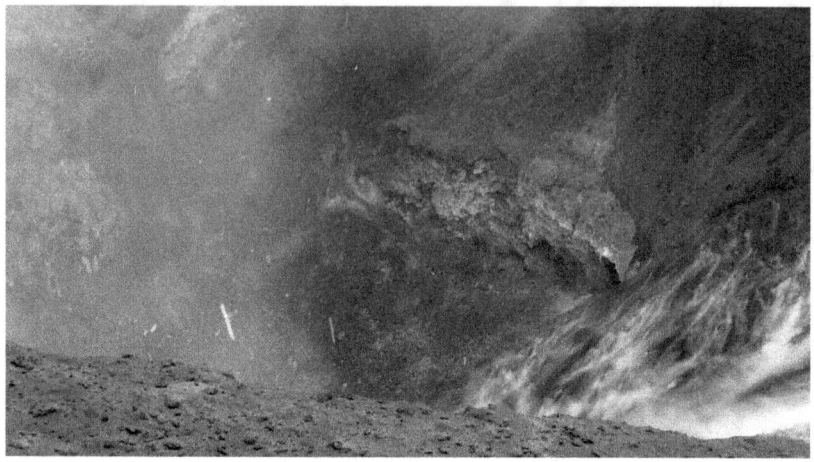

Yasur, Vanuatu.

The start of the universe in 200 words.

Let's begin with the Big Picture, known grandly as cosmology, but really more like "this is our best guess for now about what happened". Any cosmology is a bunch of theories about the origin, nature, structure, and evolution of the universe.

A cosmology is any model representing the observed universe. Western cosmology is entirely scientific in its approach, and has produced two famous models in modern times: the 'Big Bang' and the 'steady state' hypotheses.

We accept the big bang for the moment, because it predicts that there would have been five light-element nuclei formed at the big bang: hydrogen-1 and hydrogen-2, helium-3 and helium-4, and lithium-7.

The big bang theory says hydrogen-1 would be about 76% of the universe, helium-4 would be around 24%, and the other elements would be present only as traces. This is the result we get, and that offers support for a big bang, and more importantly, for a hot big bang.

We may one day see a return of the steady-state theory in a new guise, or the complete replacement of the big bang theory by something entirely unthought-of, but whatever comes next, it will still have to fit in with the laws of physics.

The most incomprehensible thing about the world is that it is comprehensible.
—Albert Einstein (1879–1955)

> My own suspicion is that the universe is not only queerer than we suppose, but queerer than we can suppose.
> —J. B. S. Haldane (1892–1964)

Forming the solar system.

> I agree. But I wonder what it would have looked like if the sun *had* been circling the earth.
> —Ludwig Wittgenstein (1889–1951), on being told how foolish the ancients were for accepting the Ptolemaic system.

All scientists agree our matter and energy must come from the stars: we are made of very ancient star stuff, as Carl Sagan said, because all the heavier atoms in the universe were produced in the stars. The atoms up to iron in the Periodic Table were formed by fusion reactions in stars, while the heavier atoms were formed by fusion reactions as stars came apart. These atoms swirled around as dust and gas.

There are several models for the dust and gas combining to make our solar system planets. To work, any model has to explain the following things (and for convenience, I leave Pluto in as a planet):

1. all of the planets have orbits that are in roughly the same plane (although Mercury and Pluto are a bit less regular);
2. this plane is close to the Sun's equatorial plane as it rotates;
3. as seen from above the Earth's north pole, all of the planets orbit counter-clockwise around the Sun, in the same direction that the Sun rotates;
4. the orbits of the planets are elliptical, but very close to circular (again with a bit more difference when you examine Mercury and Pluto); and
5. the gaseous outer planets (Jupiter, Saturn, Uranus and Neptune) are much less dense than the stony inner planets (Mercury, Venus, Earth and Mars), which are much richer in iron and silicon.

The obvious first guess to make from this is that the planets were somehow 'spun off' from the Sun, where a disc of gas and dust was thrown out to a point where it could cool down and condense into larger lumps, which then formed up into the planets, but the obvious guess is not always the best.

We know our solar system has a star in the middle, and there are lumps of rock and stuff going around it. Gravity, which we will come to in chapter 13, holds eight known and recognised planets (or nine, if you still count Pluto as a planet). Add to this an ever-increasing number of satellites to those planets, the approximately 5000 asteroids, the comets, and interplanetary dust.

Each very early planet began, we imagine, as a number of rocks, banging around in space as they whirled around the sun. Each speck of dust and lump of space rock exerted a small pull on each of the other pieces, and when two ran into each other, they crashed, mashed and bashed in the silence of space. Sometimes, they stuck to each other.

Some of the bits flew off, but a lot of the energy of movement was converted to heat, so some of the bits melted and then cooled, making a larger lump. These lumps were irregular, but subtle calculations and observations reveal that when a lump exceeds the 'potato radius' which is about 150 to 300 km, it changes from a misshapen potato to something more like a ball, and each one becomes a very hot ball, as more bits come rushing in.

The inner planets are comparatively small and dense, and are made of high-temperature condensates (mainly iron and metal silicates), while the four outer planets are much larger and mainly made of low-temperature condensates (mostly gases and ices). We can't say "ice" in the singular, because the frozen stuff includes a mix of water ice, plus frozen methane and ammonia.

Most of the sun's planets are unsuitable for life as we know it: they need to be in the 'Goldilocks Zone', not too hot and not too cold, not too wet and not too dry. They also need to be in the right size range, and that is probably the biggest hurdle to finding inhabitable planets around other stars: the easiest ones to detect are the really big ones that are too large and have unfriendly gravity.

In our solar system, only our planet falls in the Goldilocks zone, though some scientists think Mars may have supported life when the Sun was younger. For now, we must turn to planets around other stars to seek for ET's home.

In late February 2017, NASA announced that a star called TRAPPIST-1, 40 light years away, has seven rocky Earth-sized planets, all of which may hold water. Three of the planets are in the 'Goldilocks zone'. Do we have neighbours living there? As in so much of science, more research is needed.

Any Web search on the string <*physics stamp collecting*> will reveal a wealth of variants, all attributed to Lord Rutherford. It is unproven but quite likely Rutherford did have and hold—and possibly even expressed in some form—the opinion that there are two kinds of science: physics and stamp collecting, but that was in the 20th century (and just for the record, physicist Rutherford won a Nobel Prize—in chemistry!).

Among the sincere forms of philately that are undoubtedly science-as-we-know-it, two subjects are more closely related to story-telling than to rigorous proofs. These branches have little to do with the sort of exciting gadgets we see in the movies: things with dials and levers, spewing sparks, noises and fumes, and screens that pump out rock-solid results. Science rarely uses stuff like that.

The first of these two maverick subjects is evolutionary biology which identifies evidence and tests it as best it can. From time to time, small portions of the evolutionary process can be explored, but those parts are more often labelled 'physiology' or 'genetics'. The second area is geology—or most of it.

Trying to carry out igneous rock formation in the laboratory would demand vicious heat and monstrous crushing pressure, and the same pressures, along with huge amounts of time, would be needed to imitate the formation of sedimentary rocks. Simulating geological processes would require some very expensive engineering and a very large laboratory. It will never happen.

Sweeping statements like that, of course, have caught out many an observer of science in the past. Let us just agree that the odds are very much against it happening, so we must fall back on a detailed scrutiny of the available evidence.

Sadly, we have very little evidence about the rocks formed in the first billion years of our planet's history, because the oldest rocks have all been re-worked and re-formed. What the evidence tells us, though, is that the first rocks to form were the result of a hot ball of molten rock cooling down so some of the minerals in the rock could "freeze".

Most of us might think that temperatures somewhere above 1500°C were a long way from freezing, but in a technical sense, when a liquid cools and becomes solid, there is only one common word we can use in English, and that is *freeze*. We could say solidify, but I prefer to say freeze.

An igneous rock is, literally, a fiery rock, and I have to confess that I have always been fascinated by eruptions and the like. You might, if you wanted, call me a volcano groupie, but I only like them because they give us some idea of what the earliest rocks looked like.

One of the most counter-intuitive ideas in science is the notion that movement is a form of energy. I tried telling my teenage physics students about lighting fires by rubbing two sticks together, but unless the sticks were both matches, they didn't get it.

Still, when I asked them to rub a finger back and forth on a lab bench, they soon found that their finger got hot, as they changed kinetic energy to heat energy. From there, they jumped to tyres and brakes getting hot, and the idea was suddenly intuitive.

No, I am not lapsing into anecdotage: I am headed somewhere. Way back, well before the dinosaurs, the planets were bits of scattered dust and rock that somehow clumped together, pulling other bits to them by the force of gravity. These bits were whizzing along, and came to a sudden halt, losing all their kinetic energy, so the clump of rocks and dust got hotter and hotter, and started to melt.

So if we are going to put together a history of the world, we need to begin at the beginning, back when things like that were happening, when the world was new. And the best way to do that is to look at places where the world is being renewed today, around volcanoes, but only after we get a bit of background.

Backgrounder: stable isotopes.

The next item on the emergence of free oxygen will only make sense if you know what isotopes are and how we learn from them, and in particular, what stable isotopes are, and how they tell stories about the past.

Isotopes are atoms with the same number of protons in their nuclei, but different numbers of neutrons. Carbon, for example, has fifteen different isotopes from carbon-8 (^8C) to carbon-22 (^{22}C). Most of these are unstable, which means they are radioactive, and decay. Two natural isotopes, ^{12}C and ^{13}C are stable, so they don't decay, while ^{14}C slowly decays, with a half-life of 5730 years. That's the one used in 'carbon dating' (*alias* radiocarbon dating) which we will look at later.

Because isotopes have the same number of protons, they have the same atomic number and are atoms of the same chemical element. But because of the different number of neutrons, they differ in mass number, and they can shed light on past climatic conditions.

For example, water containing ^{18}O becomes more common in sea water when conditions are cold, because ^{16}O water is lighter, it is preferentially evaporated near the equator. If that water is blown away from the equator, it is likely to be locked up in polar or other ice caps. This relative increase in ^{18}O at the equator will then be reflected in the concentrations of the two isotopes found in marine shells and other material of organic origin.

Temperature and rainfall differences can influence the form of photosynthesis used in an area, favouring one or the other of the stable isotopes of carbon, and this may be reflected, for example, in the amounts of these isotopes in the calcium carbonate of fossilised egg shells. Other commonly studies stable isotopes include the isotopes of hydrogen and nitrogen.

The isotopes of greatest importance to us, and their natural abundances, are as listed below. For most purposes, we need higher precision than this list, but this sets the scene:

- Hydrogen: ^1H, 99.985%; ^2H (deuterium), 0.015%;

- Carbon: ^{12}C, 98.89%; ^{13}C, 1.11%;
- Nitrogen: ^{14}N, 99.63%; ^{15}N, 0.37%;
- Oxygen: ^{16}O, 99.759%, ^{17}O, 0.037%, ^{18}O, 0.204%
- Sulfur: ^{32}S, 95.00%; ^{33}S, 0.76%; ^{34}S, 4.22%; ^{36}S, 0.014%.

We will come to some of the methods of using unstable isotopes in chapter 14. The stable isotopes can be separated in all sorts of ways, and heavy water can be obtained by repeatedly boiling water, because what remains in the kettle is richer in deuterium oxide than the water that came from the tap. Now let's look at what some sulfur isotopes have been made to reveal.

The emergence of oxygen.

Our planet changed when oxygen began to appear. To the life that existed before oxygen, this new gas was a poison, produced by primitive blue-green algae, which are not algae at all. Technically, they are *photosynthetic moneran microorganisms* with an alga-like biology but a bacterium-like organization. They are also known as the Cyanobacteria, but they aren't bacteria, either.

The fossil record is spotty, so we know little about the period from 3.9 billion years ago, to 2.2 billion years ago, but the ratios of the sulfur isotopes in sulfide and sulfate minerals in the rocks tell us the atmosphere 2.45 billion years ago had very little free oxygen and was the main place for chemical reactions involving sulfur. Today, the atmosphere has a great deal more free oxygen, and life has evolved that depends on the once-poisonous oxygen.

We get hints of what happened from the signatures for three stable isotopes of sulfur, ^{33}S, ^{34}S and ^{36}S in sulfides and sulfates in Precambrian rocks. If you want full details, look up James Farquhar, Huiming Bao and Mark Thiemens. Briefly, they reported in 2000 that they could see a change in the earth's sulfur cycle that happened between 2090 and 2450 million years ago (mya).

Now we get technical: skip the rest of this paragraph, if you wish. Before 2450 mya, they said, the sulfur cycle was influenced by gas-phase atmospheric reactions, which also played a role in determining the oxidation state of sulfur. These reactions would only arise if the partial pressures of atmospheric oxygen were low, and that suggests to us that the roles of oxidative weathering and of microbial oxidation and the reduction of sulfur were minimal.

In short, there was a major change in the chemical reactions involving sulfur and oxygen in the atmosphere in the period between 2.1 billion and 2.5 billion years ago (bya), and during this time, the oxygen levels in the atmosphere increased sharply.

This change is recorded as a geochemical indicator in rocks of this period and older, and as it originated in the atmosphere it is clearly a global signature, a marker which was found all over the world. Around 2.2 bya, banded iron formations showed up, and we take these as indicators of increased oxygen levels that turned soluble iron (II), or ferrous or Fe^{2+} ions into insoluble iron (III), or ferric or Fe^{3+} forms.

The Cyanobacteria had been around since 3.5 bya, but now they produced enough oxygen by photosynthesis to oxidize the iron in the rocks in a process akin to rusting. There may also have been some chemical separation of water molecules into oxygen and hydrogen.

This sharp rise in oxygen levels was probably tied to the development of the Earth's ozone layer, a key element in evolution because it allowed the expansion of terrestrial life by shielding organisms from the most damaging effects of ultraviolet radiation.

From Fireball Earth to a volcanic planet.

In the 19th century, subtle thinkers asked what would happen if you had a ball of rock, as hot as the sun, floating in space? The answer was that within 20 million years, or 100 million years if you crossed your fingers and your eyes while you did the sums (few of them did that), the ball of rock would have radiated away all of its heat and be cold and lifeless. So, said most of the scientists, the world must be less than 20 million years old.

What they were missing (because nobody knew about radioactivity just yet) was that there were quite a lot of radioactive atoms inside the planet, and as these atoms decayed, they released heat, and that kept much of the centre so hot that it was liquid. Sometimes, when magma gets close enough to the surface of the planet to force its way up, and when the outside cracks, stuff oozes out, and we call it a volcano. The molten rock that erupts onto the surface is called lava.

Left, Mount Yasur, Tanna Island, Vanuatu, venting steam; and right, Yasur getting frisky.

When I was very young, the places where volcanoes were common set us a bit of a puzzle. They seemed to occur in clumps, but nobody knew why they grouped like that, until the 1960s, when plate tectonics became accepted, and then people could see that volcanoes occurred mainly in places where plates were moving against each other.

The larger plates are thousands of kilometres across, huge masses of solid rock, tens of kilometres deep, floating on the hot, viscous rock that lies deeper down. The deeper layers, we believe, are churning around, and when a plume of hot rising liquid hits the plate and spreads out, it carries the plates along, slowly and ponderously. (The Australian plate is fast, travelling ~70 mm each year.)

As the idea of tectonic plates took hold, a Canadian named J. Tuzo Wilson looked at the placement of volcanoes in 1963, and came up with the "hotspot" theory to explain the non-plate-edge volcanoes. There were places, he said, where there was a long history of active volcanoes, and this suggested that there were deep and very hot plumes that rose slowly, and could drill a hole though the planet's upper layers.

This tied in neatly with the tectonic plates idea that was rapidly gaining favour, and in particular, with the way the volcanoes of the Hawaiian chain get progressively older and become more eroded the farther northwest they are. On Kauai, there are volcanic rocks about 5.5 million years old, and that is the most north-westerly inhabited Hawaiian island.

The Hawaiian islands, moving north-west across the hotspot.

Down in the south-east, the "Big Island" is assumed to be over the hotspot, the oldest volcanic rocks on the island are just 700,000 years old, and there are active volcanoes there as well.

The original Hawaiians were aware of the differences between the islands in terms of erosion, soil formation, and vegetation. They wrapped it up in the legends of Pele, the volcano goddess. Pele had to leave Kauai when her sister, Namakaokahai, attacked her, and she fled to Oahu, then to Maui, and Hawaii (the "big island").

Today, in legend, Pele lives in the Halemaumau Crater at the top of Kilauea. Looking at it scientifically, the story tells of the struggle of volcanic islands to emerge from the sea, only to be broken down later by wave action.

So the chain of Hawaiian islands represented the northwest movement of the Pacific Plate over a hotspot that is at present located below the "big island". The heat of the plume drills through every so often, forming a new volcano, southeast of the previous one.

The hot spot model is an attractive theory, and it accounts for the vulcanism in Yellowstone, Iceland, the Azores, the Galapagos Islands and others, but in recent times, people have been looking at the facts again, and they don't *quite* stack up. There is a lot of debate going on among geologists about hotspot theory. It is possible that the end result will be that some of the hotspots will be accepted as real, while others will be explained in some other way.

In 2015, a paper in *Nature* described an Australian chain of 15 volcanoes running from Cape Hillsborough near the Whitsundays in Queensland to Cosgrove in Victoria, a chain created over more than 2000 km, over a period of 33 million years.

That hotspot is thought to be now under Bass Strait. Rhodri Davies, the lead author, said the volcanoes were discontinuous, because in places, the lithosphere, the solid outside of the planet, was more than 130 km deep, and too thick for the plume to melt its way through.

Volcanoes produce more than lava: they also emit immense clouds of ash and dust which can travel long distances, and large amounts of noxious gases, some produce different mixes of lava, bombs and ash. It all depends on the chemistry of the melted rock, the magma, its temperature, and the amount of gas, as the geochemistry influences the type of eruption. Volcanoes are also influenced by the temperature and the pressure that can build up.

Classifying volcanoes.

Geologists recognise four sorts of volcanoes:
* cinder cone volcanoes;
* shield volcanoes;
* lava domes; and
* composite cone volcanoes (also called stratovolcanoes).

Cinder cone volcanoes are the least interesting because they are steep conical hills of "cinders", meaning ash or rock fragments, but towards the end, as the gas pressure driving them eases, lava often flows out and covers them. The ash they produce can cover quite a large area, wiping out all plant life and either killing or driving out animal life, as Parícutin did in Mexico in the 1940s and 1950s.

Inside the crater of Hverfjall, a cinder cone volcano in Iceland.

Lava domes are comparatively rare, but they produce a lava that is much more viscous, less runny. The lava gloops out and sits there, piled up around the vent. There is a catch, though, when the vent is blocked by the first lava to come out, as happened with Mont Pelée in Martinique in 1902. There was a massive eruption there, and *nuées ardentes* (French for "glowing clouds") of superheated dust and ash rushed down on the town of St Pierre, killing 30,000 people.

Composite cone volcanoes are the ones I used to love drawing diagrams of when I was a small boy: they have a mix of layers, some of them lava, some of them made up of "volcanic ash" (small rock fragments), and that tells us straight away that these are much more furious.

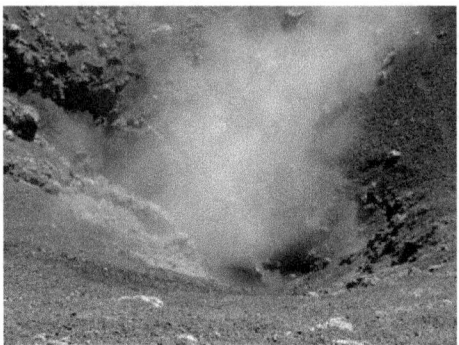
The crater of Mount Etna on Sicily, a far from extinct composite cone volcano.

Krakatoa, Vesuvius, Mount Etna, Fujiyama, Taranaki and Mount St Helens are just some of the better-known examples that fit into this category.

Taranaki, a composite cone volcano on New Zealand's North Island. The clouds are just clouds.

Mt St Helens in the USA, once a neat cone, until the side nearest to the camera blew out.

Shield volcanoes are the longest and widest, and may be tens of kilometres from side to side, though Rangitoto, just off Auckland (below), is a great deal smaller. They have shallow slopes, and all the Hawaiian islands were once active shield volcanos.

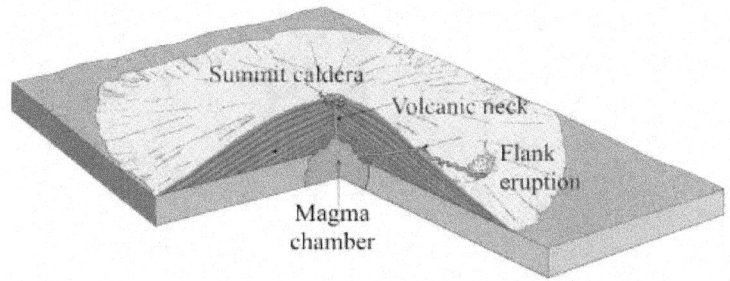

Diagram of a shield volcano.

The "big island" of Hawaii has Mauna Loa, a shield volcano that is the world's biggest mountain, because from sea floor to summit, it is more than 9 kilometres high. It stretches for more than 150 km from side to side. Shield volcanoes pump out low-viscosity basalt, which flows a long way, which explains the low, flat shape.

Rangitoto, a dormant (for now, at least) shield volcano, off the eastern suburbs of Auckland, New Zealand.

The shield volcanoes sometimes open a side vent, and when they do, huge amounts of basalt can flood out and rush across the countryside, producing flows, which we will look at later.

Geysers and vents.

Volcanic areas often have geysers, where groundwater is heated under pressure until it boils, pushing out overlying water, which then boils explosively.

It wasn't a geologist who showed that volcanic rocks are pretty much the same everywhere. It was a chemist, and he wasn't just any chemist, either, but Robert Bunsen (1811–1899), who made many other inventions and discoveries.

These inventions do *not* include the Bunsen burner, which was actually invented by Peter Desaga and popularised by Bunsen, but the list does include the Bunsen grease-spot photometer, and a model of a geyser which can often be seen in superior science museums. This is made up of a flask full of water, a long tube coming up out of the flask, with a shallow pan at the top.

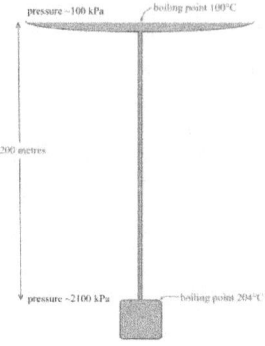

Bunsen's model geyser in its modern form, but with real-life dimensions and values.

In the 1840s, Bunsen visited Iceland to study the active volcano Hekla, but he also visited Geysir and wondered how these 'fountains' of water and steam operated. What made the fountains keep on spouting, usually at regular intervals? At Yellowstone, 'Old Faithful' was so accurate that even its name celebrated its impressive time keeping. We will begin with the simple version:

Two views of Old Faithful, Yellowstone USA.

Bunsen knew two things—that the boiling point of water depended on the pressure acting on it (extra pressure means a higher boiling point), and that the geysers were supplied with a great deal of heat, coming from deep below. From that, he designed a simple model to show how they work.

Bunsen assumed a large, deep open chamber, with a narrow shaft connecting it to the surface. The model needs one extra thing: a source of heat like a Bunsen burner and, while the diagram shows a 200-metre tube, even the two-metre-high model works. (The short model is safer and a Bunsen burner can drive it!)

We start with the open chamber full of water, just after an 'eruption' of the fountain. The water, now cooled below 100°C, has trickled down from a surface pool into the empty chamber. Water fills the chamber and the whole tube, so the water in the chamber is under a pressure of 2100 kilopascals (21 atmospheres), and at that pressure water only boils when it gets to 204°C.

The hot rocks heat the water up and when the temperature reaches 204°C, some of the water at the bottom changes to steam, some water is pushed out of the top, and with less water in the tube, the pressure drops.

Now the superheated water boils furiously, pushing more water up the tube, lowering the pressure, and making the boiling even more furious. The result is an 'eruption' of hot water. The water spreads out across the surface in a large pool, the hot steam rushes out, and then water in the pool flows back down, filling the chamber with cooler water and the process starts again.

This is not a perfect theory, but the experts say it will do until a better one comes along. Now let's take a more detailed look at the geyser effect.

To understand geysers, we need to look at the physics of *boiling*, because water doesn't always boil at 100°C. Even cold water and ice give off water vapour, but when you heat water, it gives off more and more vapour, and the pressure of the water vapour increases. Once the vapour pressure of the hot water is greater than the air pressure outside, bubbles form, and we say the water is boiling.

When the air pressure is "one standard atmosphere", or 101.3 kilopascals, the water boils at 100°C. If the pressure is reduced to 50 kPa, the boiling point is just 78°C, and at 6.3 kPa, the boiling point is just 37°C, or body temperature. At that atmospheric pressure (which occurs about 18 km above sea level), you not only can't breathe: you will begin to boil!

The average atmospheric pressure at the top of Mount Everest, the atmospheric pressure is about 34 kPa, so the boiling point of water is 71°C. As my old physics teacher told us, you can't get a decent cup of tea when you go up a mountain. On Mars, the pressure is about 0.6 kPa, and tea of any sort is impossible to get, but that's the least of your worries.

Thermal pools, Orakei Korako, New Zealand.

In the Bunsen model, cooler water from the pan trickles down into the flask, and the whole thing starts again, just like Old Faithful. A real geyser is the same, but on a larger scale. There are differences though, because the model is made of metal, and a real geyser is made of rocks, which slowly dissolve.

Thermal pools, Yellowstone area, USA.

Thermal pools, Fountain Paint Pots, USA.

Mineral deposits, Mammoth Springs, USA.

Boiling mud spring, Yellowstone.

Volcanic chemistry and physics.

Nobody much knows Bunsen invented the model geyser, and only a few specialist historians now recall that he was also the founder of geochemistry.

> It cannot therefore be doubted that the extensive volcanic elevations constituting the high table-land of Armenia and the island Iceland have flowed from sources which were *chemically* identical.
> —Robert Bunsen, *Poggendorff's Annalen*, 1851.

Bunsen's careful chemical analysis showed that his sample rocks, collected in both Armenia and Iceland, were identical, and so he invented geochemistry. He was lucky, because different sorts of eruption can be very different in chemical terms, and he must have chosen two volcanoes of the same kind.

Some volcanoes make pumice like the Kermadec islands one that made my beach visitors; some make viscous and gloopy lava like Kilauea; and those like Mt Yasur (at the start of this chapter and again, below) have explosive eruptions, throwing lumps of molten rock high in the air, where they turn solid and fall as "volcanic bombs".

Volcanic bombs are defined as solid lumps of more than 64 mm diameter that are ejected during a volcanic eruption. Some of them can be up to several metres in size. (Geologists in America who have these in their carry-on luggage call them "volcanic rocks" because TSA agents are not very bright.)

Those that are highly liquid when they are launched can be deformed in flight or when they hit the ground. Experts recognise a number of curious classes including breadcrust bombs with surface cracks, cored bombs, 'cow-pie' bombs which flattened on impact and others.

Volcanic bombs on Mount Vesuvius. The right-hand one was about 800 mm diameter.

Staying around when the bombs start to fall is a no-brainer (in several senses).

Parasitic cones.

In 1835, Charles Lyell wrote at length about the many "minor cones" found around Mount Etna on Sicily. These are often called parasitic cones, but under either name, they refer to lava finding an alternative route to the surface.

Charles Lyell, *Principles of geology*, volume 2, "Minor cones on the flanks of Etna".

My main reason for raising the topic here is that I believe I recently found the planed-off remnants of a parasitic cone in the tidal waters of Waitemata Harbour, off the eastern suburbs of Auckland in New Zealand.

Rangitoto and Waitemata Harbour, Auckland, New Zealand, at high and low tide.

At high tide, there is no sign of it, but when the tide goes out, a wide rock platform is exposed, and if you look at it from the wrong angle (as in the left-hand picture below), there is nothing of interest to see. In the right-hand picture below, looking at the same rock platform, something different emerges. To really see what is going on, you need to look for concentrically curved dipping beds.

Parasitic cone remnants (?), eastern end of Mission Bay, Auckland, New Zealand.

Earlier in this book, I showed an eroded cross-section of an old volcano near Cape Palliser in New Zealand, which showed a series of dipping beds from a composite cone volcano. Here it is again, in miniature.

That old volcano near Cape Palliser, North Island, New Zealand.

My inexpert but logical interpretation is that above, we see the planed-off remnant of an ex-composite cone volcano. As the careful reader may have guessed, I am something of a vulcanophile, but I only venture onto the outside at a time when the volcano is relatively safe. Going inside a volcano, as we will see shortly, must be even more carefully judged. First, let us consider external visits.

Visiting a volcano.

One of the most remarkable eruptions ever was called Thera, though the remnants are now called Santorini. Thera erupted many times, but the Big One happened around 1628 BCE, going on tree ring studies in Ireland and ice cores. That year, the atmosphere was full of dust, leaving deposits in the ice and stunting the growth of trees in many places by blocking out the light.

Thera probably threw 60 km^3 of rock and dust into the atmosphere, around four times as much as the famous Krakatoa eruption. It probably gave us the legend of Atlantis, and destroyed Knossos, ending the Minoan civilization.

Santorini most recently went off in 1950, and came close to eruption in 2012, when, according to informed gossip, the Greek government decided to say nothing about the looming threat, because the island was a major tourist draw-card. It all turned out well when the risk went away, but one of these days, it will go again, probably trapping thousands of tourists.

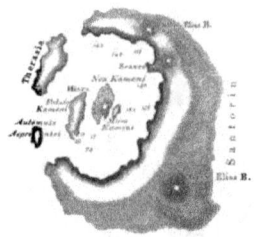

Map of Santorini, a blasted-out crater, Greece, from Lyell's 'Principles'.

Most tourists come on cruise ships, but others come by inter-island ferries, staying on shore and walking the tracks that skirt the rim of the crater remnant. As a rule, dormant or inactive volcanoes are the best to walk on, because there is less of a sulfur smell about them, and there is less risk of frizzling.

Two views of Santorini, Greece.

Extinct volcanoes are the best bet of all, as they are starting to wear away, which means you don't have to go quite so high—and height can be a problem. If you ever go up Mauna Kea, which has been quiet for about 4500 years, you will probably go by bus.

Warning sign at Hale Pohaku, Mauna Kea.

This is less wearing than walking, but leaves you at risk of "altitude sickness", because the summit is 4200 metres above sea level Altitude sickness typically shows up as dizziness, and sensible tourist drivers pause at a convenient souvenir shop, and watch you like a hawk.

The only time I ever had any altitude problems was on Mt Etna (3300 m), where we were rushed up the mountain in a tourist coach, bundled into a cable car, and whizzed to the summit, with no checks at all. Feeling mild symptoms of altitude sickness, I just took it quietly, sauntered around, and enjoyed the views.

Because it is so high, usually above the clouds, astronomers like Mauna Kea.

Mauna Loa still goes off from time to time, with 33 eruptions since 1843, and Kilauea was dribbling out lava when I was there in 2005. It was a 6 km walk out in twilight and a longer walk back (the guide got a bit lost) by flashlight. As I write this, Kilauea is far less safe.

Even back then, the lava was fresh, and at the furthest point on our stroll, as the night sky started gathering stars, the hillside above us seemed to gather stars as well. These were the first glimmers of glowing lava, slurping down the hill.

You can't take a flash photo, so this is the best you will get of the first glows on Kilauea.

What we were seeing was the glow of red-hot lava emerging from a crust of solid lava, and then slipping back in again, running down what geologists call lava tubes (we will come to those shortly). Later, we were able to capture more convincing shots of the lava, working its way down the hill.

Lava flow on Kilauea.

So long as you are well away from the sea, where the lava drops over a cliff and generates acidic steam, the atmosphere is fine. You are close to sea level, so there is plenty of oxygen, but there are other nasties as well.

Warning sign near sea level on the flank of Kilauea: the hot lava runs into the sea and creates all sorts of unpleasantness.

The lava is brand new and remarkably sharp, so you need to watch where you are going, and you need to watch the wind direction. You could go out on your own when I was there, but since even an experienced guide can get lost, this is not a good idea.

Mount Yasur, Vanuatu: after dusk, the flying lava shows up well. The night after we were there, bombs were falling on the observation point, and people had to move.

If you want a gently active volcano, Mt Yasur on the island of Tanna in Vanuatu is a good choice. There will be a few bombs and lots of fumes, but mostly, you will be safe enough. The only way to get there is with guides who know the place, but there is usually more fun to be had looking at the aftermaths of eruptions.

Craters at Mt Vesuvius (left) and Mt Etna (right).

Lava flow surrounding a house at Mt Etna (left) and covering a road on Kilauea (right).

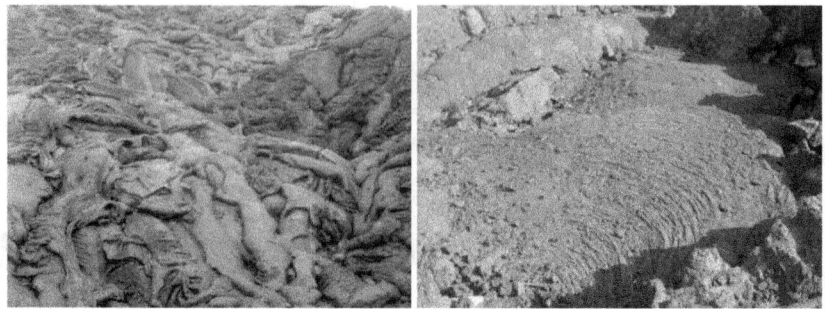
Lava flow on Kilauea (left) and Mauna Kea (right).

Volcanic necks: Beacon Rock, Columbia River (left) and Bartolomé, Galapagos Islands (right).

Two views of what appear to be the flanks of extinct composite cone volcanoes, Columbia River, USA.

Two views of a scree in the crater of Mt Vesuvius.

Detritus, Mt Etna (left) and a section through ash deposits, Bartolomé, Galapagos Is.

Going inside a volcano

Thrihnukagigur (*Þríhnúkagígur* in Icelandic, meaning 'three peaks crater') has been dormant for 4000 years, and it has been open to tourists since 2012. Going there is definitely a summer activity, because to visit it, we had to walk 3 km over a lava field in mist, while biting winds drove chilly rain into our faces. This was summer in Iceland, so it could have been worse, and at least it was warmer in the magma chamber, at 6°C. At the time, that *felt* warm!

Access is now by an open cable lift, which you can see below. It lowers eight people at a time, down through a small surface vent, onto the floor of the magma chamber 120 metres below. The chamber's first explorer, Dr Árni B Stefánsson, was lowered down on a rope in 1974, in far less comfort.

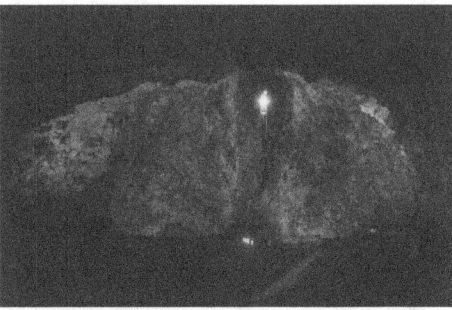

This cable lift brings visitors down into the magma chamber, Thrihnukagigur, Iceland. (Courtesy Benjamin Hardman, *Inside the Volcano*)

When Thrihnukagigur went quiet, lava stopped pouring out, and the top of the magma sank, leaving a gap, 120 metres deep, below the tiny vent on the surface. Then the magma 'froze', and the floor formed on top of it. The rocks on the side of the chamber, and also on the floor, are large, angular and jagged.

The walls of the magma chamber, Thrihnukagigur volcano, Iceland.

This trip is suitable for moderately fit people between 12 and 75 who can walk a fairly level (but chilly) path for 3 km. It ought to be on any rock hound's bucket list. Bookings can be made online, and buses leave from Reykjavik.

Lava tubes.

These are also called lava tunnels, and they form when lava flows over a rocky surface, usually down the side of a volcano. Because air is much cooler than lava, the top of a flow forms a crust while the lava keeps on running through underneath, producing a conduit or tube.

Most of the time, the tubes form in the more gloopy *pahoehoe* lava, but they have also been known in the less gassy, more crystalline, more jagged *a'a* flows on Mt Etna. The most interesting thing is the way the tubes carry molten basalt a long distance from its source before it cools down, because the walls of the tube act as an insulator. Most tubes are less than 1 km in length, but one lava tube in Queensland is more than 100 km long!

Large lava tube, Santa Cruz, Galapagos Is.

The lava is typically ~1150°C, which makes it very fluid and fast-moving. One US source records the speed of flow in a tunnel on Mauna Loa in 1984 as 35 mph (50 km/h near enough). At that speed, the tunnels stay open because there is *thermal erosion* in the inside of the tube, as the hot lava melts the walls and carries the remelted rock away.

When the eruption stops, the remaining lava drains out, leaving an empty tunnel, ready to be explored, if it is large enough.

Small lava tube, Santa Cruz, Galapagos Is (left) and what *appears* to be a lava tube in the bank of the Columbia River, near The Dalles, Oregon: I am keeping an open mind about that one.

Life on a volcano.

A volcano is bad for life: it burns animals and plants, choking them with noxious gases, and for many years after the volcano settles down, there is no water to be had.

Eventually, a few seeds blow in on the wind. On volcanic islands, seeds are unlikely to be brought in by birds, because there is no food for birds there at first, so wind-blown seeds are the usual thing.

This plant was much photographed, but nobody seemed to have known its name when I was there. It must have come from a desert environment. Bartolomé, Galapagos Is.

Things are a bit easier for adventurous life forms where the volcano is surrounded by areas that are still populated. The pioneer plants still need to be able to withstand dry conditions, and because any animals in the area will be very hungry, the plants need to be tough and either spiny or bad to eat. For the most part, the plants settle in sheltered spots.

Tough plants on Mauna Kea, Hawaii.

Over time, more plants arrive, taking advantage of the shelter provided by the pioneers—and probably also using the thin soil the pioneers' roots have managed to hold in place.

On Mauna Kea, the biggest danger to the plants comes from clumsy tourists. As tourists are unnatural, the authorities use unnatural (de)fences.

Soon, there is a small community of plants, and the first insects can move in to eat the plants. Once a few insects have arrived, birds and reptiles can settle, and have some hope of surviving. Now life has a hold once again—and it refuses to let go.

A lava lizard eating a grasshopper, Bartolomé, Galapagos Is.(left) and a ladybird, Mt Etna (right).

Basalt flows.

Basalt is mainly made up of the mineral pyroxene, which is rich in iron and magnesium, together with plagioclase, a calcium-rich feldspar. Given the way the minerals are finely mixed, some geologists just prefer to speak of the percentages of different elements in the igneous rocks.

Using in that system, granite is about 72% silica and 13% aluminium oxide, while basalt is 50% silica, 16% aluminium oxide and basalt has far more calcium and magnesium compounds. Moving away from technicalities, basalt is easy to spot because it is darker, and because of some of the forms that it takes.

First, we will look at the way basalt sometimes forms *traps*, otherwise known as flood basalt. By the way, 'trap' in this case comes from a Scandinavian word *trappa*, meaning stairs. When you look at the pictures, you will see why.

Flood basalt flows in the Columbia River Basalt Group, USA, with "steps and stairs" effect.

It appears that huge volumes of basalt were pumped out, and while the first few layers would have filled in the wrinkles and bumps in the landscape, the later flows just ran and ran, across the flat plain created by earlier flows.

Basalt flows fill in the shape of the rock they cover. Astoria area, Columbia River, USA.

The typical step-like pattern of flood basalt, after weathering and erosion. Columbia River, USA.

The main thing that distinguishes basalt is that when it cools, it often forms columns, and these are known from many parts of the world.

Columnar basalt.

I mentioned columnar basalt at the end of my introduction, when I described how I joked about demanding that the US government destroy certain deposits on the Snake River. I said the deposits were contrary to the laws of physics. To see why I reacted this way, we need to consider just how columnar jointing arises, and then I will explain why I was wrong to be surprised.

Curved columnar basalt on the Snake River.

Typically, we see columnar jointing in lava flows, though it can also be encountered in basaltic sills and dykes—and as I will show later, it can arise from contact metamorphism when sandstone changes to quartzite. For now, we will stay with the lava flows.

Straight columnar basalt in Fingal's Cave, Staffa, Scotland; Svartifoss Iceland.

When a lava flow cools, it shrinks, and especially in the middle section of the flow, straight-sided cracks that geologists call joints, begin to form. The process is a little like the way mud will crack as it dries, but that can be a misleading analogy.

The joints are somewhat regular, and they shape the rock into columns that are, on average, hexagonal, but the number of faces can be 5–7 in Fingal's Cave in Scotland, or 4 to 8, or even 3 to 12.

As a rule, the columns are vertical, but this is subject to an important condition: the column is typically perpendicular to the contact, which means the direction of cooling. Charles Lyell wrote this:

> [Assuming] that columnar trap has consolidated from a fluid state, the prisms are said to be always at right angles to the cooling surfaces...
> —Charles Lyell, *Elements of Geology*, 183, 1838.

Having explained that curved columns can happen easily enough, two pages on, he offered another example from a valley north of Vicenza in Italy, and this time he illustrated it as well.

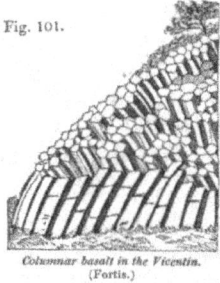

Curved columnar basalt in Italy, from Charles Lyell's *Elements of Geology* (1838), 105.

So before you call in the army engineers to right a scientific wrong, think carefully: science just is, and when the evidence goes against what you believe, don't attack the evidence: attack your beliefs, and work out where you went wrong! Now here are some other images of columnar basalt:

Curved and straight columnar basalt on the Snake River.

Two views of straight columnar basalt in Fingal's Cave, Staffa, Scotland.

Dykes.

There are many famous dykes in Britain that have nothing to do with geology. They are earthworks that date back to the Dark Ages, and the most famous one is Offa's Dyke (*Clawdd Offa* in Welsh), which runs roughly along the border between England and Wales. This is often explained as a defensive wall like Hadrian's Wall, but historians say it was really just a clear boundary.

Credit for building it is given to a Saxon king of Mercia, but at least some parts of it have been shown by carbon dating to be older, and there is no record of garrisons ever being stationed on the dyke. A dubious tradition has it that the English cut the ears off any Welshman found east of it, while the Welsh would hang any Englishman found to the west.

The point about earthworks is that they are both a ditch *and* a mound, formed from the earth that was dug out of the ditch. In the English language a dyke (or dike) can be the sort of sea-wall that the little Dutch boy stuck his finger in, or a drainage ditch. It is in my nature to diverge, but this linguistic divergence is an essential one, so please bear with me for a moment.

The first reference I can find to a dyke in the geological sense comes from a review of a Scotsman's book in *The Edinburgh Review, Or Critical Journal*, in 1818.

Basil Hall was a captain in the Royal Navy, and he undertook a voyage to "…the west coast of Corea and the great Loo-Choo island". If we take Corea to be Korea, and the Loo-Choo islands are the Ryukyus, then the great island is probably Okinawa. That is background: while there, he saw "…a whin dyke, four feet wide, the planes of whose sides lie N.E. and S. W…" off the coast of Korea, but he drew something far better at Madeira.

Dyke near Madeira, from Charles Lyell's *Elements of Geology* (1838), 105.

This was a classic example of a wall-type geological dyke, and Charles Lyell used it in his *Elements of Geology*, but he had an even better one that he had already used in the third volume of his *Principles of Geology*, a generation earlier.

The engraving, seen below) showed dykes at the base of Mount Etna on Sicily, and like the Madeira example, the stone of the dyke is longer-lasting than the layers of volcanic ash that it pushed into.

Dyke near Mt Etna, from Charles Lyell's *Principles of Geology* (1834), volume 3.

There are other dykes that weather out quickly, leaving an empty ditch, but we will come to those in a moment. In 1824, Alexander von Humboldt wrote (in translation) in the *Edinburgh Philosophical Journal*: "At Skeen in Norway, a basaltic and porous syenite…is a bed, not a dike…"

Clearly, Humboldt meant to draw a contrast between horizontal and vertical layers of rock, but we are looking at a translation written by educated members of the Scottish Enlightenment, who knew all about dykes, but at that point, most of the geologists were Scots, and they knew what Scottish dykes were.

Dyke on the flank of Mt Vesuvius.

When Lyell and Roderick Murchison, later to be two of Britain's most famous geologists, wrote about a lava flow in the area of Puy de la Vache in France, they referred to it being "…in the shape of a long dike or mound, rising in the middle between 100 and 200 feet high…" We would not call that a dyke now.

Yellowstone River, USA: from the appearance as seen from a distance, there are probably dykes in the valley that intruded into volcanic ash.

The word "dyke" came to mean a fairly narrow seepage of molten rock through a crack or a crevice in ash or rock. The picture above shows the Yellowstone River valley, and it appears to be a valley through volcanic ash, with several dykes exposed near the bottom. The dark band at the top looks like a later basalt flow.

Where the stone that makes the dyke breaks down quickly, instead of a wall, we have a ditch left behind. The picture below shows a weathered-out dyke on the Australian coast between Gerringong and Kiama.

Weathered dyke between Gerringong and Kiama, NSW, Australia.

In this case, the effect is exaggerated, because the dyke magma came up through a joint in sandstone. The magma was hot enough to cause what we call contact metamorphism, which we will look at in chapter 6, but for now, think of it as a way of toughening the rock.

Here are two more examples of dykes, both from the south coast of New South Wales.

Dykes showing variable weathering, south coast of NSW.

2: Deep rocks.

Temperature gradients in ordinary [volcanically] quiet areas range from less than 10 to as much as 50 degrees Celsius per kilometre.
—A. E. Benfield, 'The Earth's Heat', *Scientific American Reader* (1953), page 71.

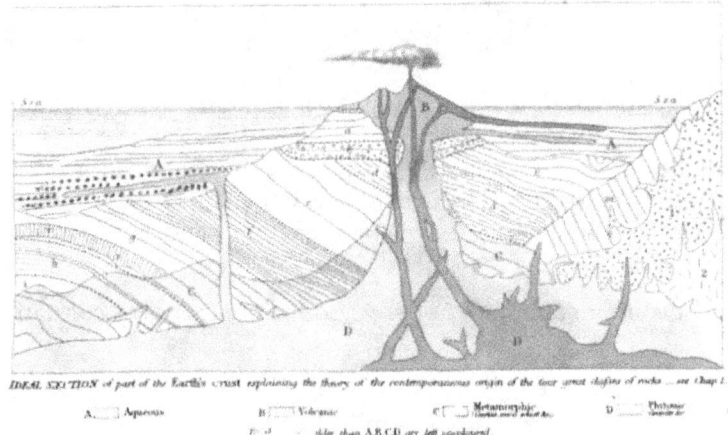

A view of the 'inner earth', the frontispiece from Charles Lyell's *Elements of Geology* (1838).

In his 1838 *Elements of Geology*, Charles Lyell's frontispiece (above) identifies four types of rock: aqueous (we call them sedimentary now); volcanic; metamorphic and plutonic (from Pluto, the Greek god of the underworld).

Now we combine plutonic and volcanic as the igneous rocks, but *plutonic* is still used to mean the large-crystal igneous rocks, not that it matters much, because most of the plutonic environment is beyond our view, and as such, beyond the scope of this book, which is about rocks we may hope to meet. This chapter begins with Plutonic Rocks Lite, but then looks a little more carefully about how we know what lies beyond our reach, and what these rocks are.

Crust, mantle, Moho and core.

Our knowledge of the inside of our planet depends on seismic waves, and a lot of mathematics. Each earthquake sends out primary and secondary or P and S waves. These go out equally in all directions, but they aren't received equally in all places, and as a rule, the further you are from the epicentre of the earthquake, the point the waves are treated as coming from, the longer they take to get there, if they arrive at all.

If a seismic station is more than 103° around the world from the epicentre, it will be in a shadow zone for S waves. There is a similar shadow zone for P waves that starts at 103°, but if the station is more than 143° around the world, it will receive P waves once again. Waves don't always travel in straight lines, which is how bays and beaches get curved (see the end of chapter 4). The curving is called refraction, and the same effect determines how lenses of all sorts—from the ones in our eyes to those in our cameras—work.

Just take that as true, or look it up, because I won't be explaining lenses here, *or* the mathematics that allow us to deduce from the seismic waves that the planet has an outer layer that we call the mantle, and a two-layered core, which the sensible geologists call the inner core and outer core.

Working outwards, the inner core lies between about 6370 and 5150 km down; the outer core between 5150 and 2980 km; the deep mantle lies between 2980 and 650 km; but above that there is a transition zone, also counted as mantle, from 650 to 400 km, and above that again is the upper mantle, from 400 up to 50 km where there are continents and 10 km where there are deep oceans.

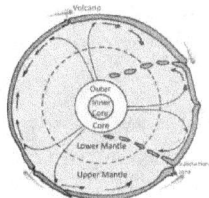

How the core mantle and crust relate to volcanoes (not to scale).

The upper limit of the mantle is variable, because the continents float on the top of the mantle, just as a floating ship has part of its hull under water. That floating is how we happen to get plate tectonics. If you were to weigh them, the crust is about 0.3% of the planet, the mantle is around 67%, the outer core is about 31%, and the inner core is around 1.7%.

The core is usually thought to be responsible for the planet's magnetic field, which does a lot more for us than make compasses point the right way: it also helps protect us from particles that come in as part of the cosmic radiation. That's probably all we need for now, except that the informed observer may have noticed that we still have to meet the Moho or Mohorovičić discontinuity.

This lies between the crust and the mantle of the Earth, and was detected by a careful analysis of shear waves from earthquakes originating at a distance of less than 800 kilometres. We now know that this layer is closest to the surface in the middle of an ocean, deeper under a continent, and deepest of all under a mountain range. You will never see it, but you need to know it's there.

Magmatic differentiation.

As we have seen, one of the standard principles of science is that things get more and more disordered. Some people might argue that it is therefore impossible for a seething, searing mineral soup to form crystals like those found in granite. Some people claim the second law of thermodynamics proves that evolution is impossible. These people are cherry-picking without paying attention, and any person who does this probably denies climate change as well.

It is the overall *average* disorder that always increases, and sometimes the achievement of disorder in one corner of the test tube leads to an increase of order in another. (In hypothetical science, square test tubes are all the rage, and often used for preparing calabash of spherical cow. Trust me, I'm a scientist!)

Sorry. We blame the way crystals form on magmatic differentiation, which simply means a magma can, as it cools, form different areas of order. Back in the 1920s and 1930s, Norman Bowen developed something called the Bowen Reaction Series, which is the order in which crystals form as magma cools.

The main thing to note is that, as you draw certain minerals out of the liquid and deposit them somewhere as crystals, the composition of the remaining magma changes. The glossary entry is a mass of jargon that will help you look this up if you wish, but this is Not Recommended. OK?

Crystals.

We all know rocks when we see them, but what's inside them? Most rocks, most ice, even most metals, are made of crystals, and crystals tell us atoms are real, because only identical objects pack neatly as crystals do. This may seem bizarre, but the simple fact is that people can do chemistry without believing in atoms. That said, explaining is much easier if we assume that atoms are real.

In the early 1900s, Albert Einstein explained 'Brownian motion', and everybody was excited, because this was the first proof that atoms were real, but about 300 years earlier, Johann Kepler published a small book about 'hexagonal snow'. He deduced from the shapes of snow crystals that snow must be made up of densely packed spheres.

Sadly, nobody really ever took much notice of this work, and Kepler's idea was forgotten, because the world was not ready for it. If you drop marbles, lead shot, ball bearings or tennis balls, down one by one on a gently sloping flat tray, so they roll down to the lower rim, the spheres will pack neatly into place, fitting snugly together, especially if you give the tray a nudge, every now and then.

In a gumball machine, the gumballs pack regukarly.

Maybe, if gumballs had been common in Kepler's time, the idea of atoms might have taken off. Things with a regular shape seem to settle down naturally into regular patterns. A number of people investigated crystals in the seventeenth century, but it was only about 200 years ago that a French botanist, a trained observer, dropped a friend's calcite crystal, and noticed something odd.

René Haüy dropped the crystal, it shattered. Picking up the pieces, he saw that even the smallest fragments had identical faces, and the angles between the faces were the same. Like Kepler, he realised that crystals must be made of smaller units in the same form as the large crystal.

As we understand it now, a crystal is a substance that has grown freely by the addition of individual particles, so that it can develop external faces. A crystal forms into a crystal system that is dictated by the shapes of its particles. Metals are crystalline in most cases, so are salts and many other solid chemicals.

The study of crystals is either crystallography to scientists, or gullibility if you believe that crystals have some magical power to cure illness. Crystals do indeed have an amazing healing property, but only for the sick wallets of crystal sellers, although they have also been used to resuscitate dying bank balances.

The only 'magical' power a scientist can get from a knowledge of crystals is that sometimes, you can identify an igneous rock by looking at the crystals in it with a hand lens. That aside, crystals are beautiful, and there's nothing wrong with collecting attractive specimens: just don't get your hopes of magic up!

Cast iron is made up of crystals, so if you think crystals can cure any illness, you should hit yourself on the head with lumps of cast iron. Maybe you already did, and that is why you have this irrational belief.

Getting serious again, if you let dissolved molecules and ions drift out of solution, they pack into neat crystalline arrays. Each particle slips into a regular pattern that repeats, over and over again, just as the gumballs did. Usually, gumballs and other spheres will pack into a triangular set-up.

Wherever you look, order seems to be normal, at least in solids, but in crystals, because the particles are often several sizes and shapes, crystals have more interesting shapes than the ones based on triangles. Table salt, sodium chloride, forms cubic crystals, and so does lead sulphide, or galena.

But even at the very best, the order is only local, and all crystals contain small imperfections. Somewhere in the array of marbles, there will have been a hiccough, a point where the pattern has been forced out of shape by one marble too many or one marble too few. Nearby, there will be a new ordered array set up, at a slight angle to the first one.

In igneous rocks, we see crystals meshing into each other, because the melted rock started to crystallise at many different places, with each crystal at a slightly different angle. The crystals are still there, but when they are blocked, they stop growing and start off in new directions.

Things got complicated after Auguste Bravais (1811–63) worked out mathematically the number of different 'bricks' which might be used to develop the different crystal houses, and invented what we now call Bravais lattices.

Now people seemed more ready for the idea, and scientists began to study crystals to see what they could learn from them. Max von Laue, a physicist, explained the delay another way. He said the physicists, who might have solved the problem sooner, never saw really good crystals. The mineralogists, who saw and hoarded the finest crystals, lacked the necessary physics to realise what secrets their specimens held.

The other notable thing was cleavage: if you try to tear a piece of woven material, you will quickly find that there are two directions in which it tears easily. Except in knitted material, cloth has two sets of threads at right angles to each other. If you try to tear along the line of just one of these sets of threads, your job will be far easier than if you try to tear diagonally across both sets at once.

Cloth is a two-dimensional lattice, with its threads in a regular pattern, weaving around each other. The properties of the cloth we make will depend on the properties of the threads, and also on the way they are arranged, or woven. A knitted cloth is less likely to tear than a woven material, but it is much more likely to run, if we pull out a single thread.

The particles in a crystal form a three-dimensional lattice, and the same effects can be seen in both crystals and cloth. In some directions, the splitting or cleavage, will be much easier. These cleavage directions are parallel to the directions which are in the lattice. Yet even if they are cleaved, crystals retain their symmetry, and when they grow, new particles are drawn in to fill the gaps in a crystal surface.

Wandering particles are more likely to stick to a gap which is filled on three sides, and once they stick, they are less likely to be knocked off again. This is why, in theory, crystals grow neatly, with flat sides, and remain symmetrical. The reality is shown by this simple activity that I developed for younger readers of a related children's book, *Australian Backyard Earth Scientist*.

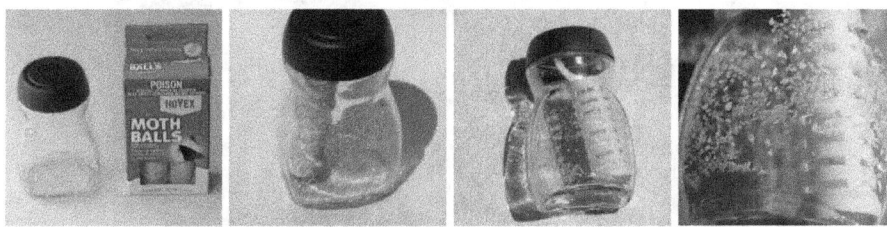

The materials on Day 1, the jar at the end of Day 1, Day 5 and Day 11.

You need mothballs, a clean jar with a tight lid, a suitable place to leave it, and two weeks or so. Put the mothballs in the jar, tighten the lid, and leave it in a safe sunny place, away from wind, pets and stray animals including younger children. R*emember that naphthalene is a household poison: don't touch, smell or taste it — and keep the lid on*! Just look and watch the crystals.

There are six crystal systems, but like Bravais lattices, these are only important to specialists.

Granite.

We only see granite on the surface if it is pushed up and uncovered. Once it is uncovered, it weathers in a particular way. (We will come to weathering in the next chapter: the main thing is that all rocks break down when they are exposed.)

Granite tors near Cathedral Rock and Mt Yarrowyck, New England region, NSW, Australia.

Remarkable Rocks, Kangaroo Island, and Middle Cathedral Rock, Yosemite California.

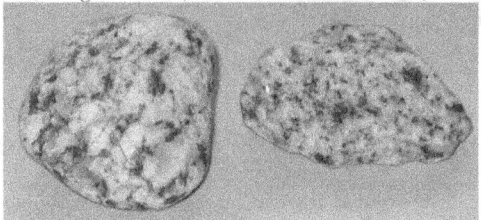
Hand specimens of granite.

And now we have the igneous rocks as a base store of rocks to look at, we can move on to the rock cycle, which is driven by two things: weathering and erosion. In that discussion, we will look more closely at how granite tors form.

3: How rocks wear away

Weathered rocks in Australia.

From rocks to soil

> In the agricultural sense soils are the superficial layers, usually less than a foot in thickness, of disintegrated and decomposed rock material, which is mingled with organic matter, and furnishes the necessary conditions and materials for plant growth.
> —G. W. Tyrrell, *The Principles of Petrology*, Methuen, 1929, p. 184.

Humans understood erosion long before anybody had the idea of weathering. In 1802, John Playfair invented geomorphology when he gave us Playfair's Law, saying that rivers cut their own valleys, rather than following pre-existing routes. He never realised that rock minerals make both physical and chemical changes as they weather to form soil, with some of the soluble products being leached away by groundwater. Let's begin with the rocks breaking down.

What we call "weathering", the decay of the rocks, combines with the many ways the planet can move rock debris (erosion in other words), to make the rock cycle operate. A lot of the most spectacular scenery emerges because some parts of the rock resist weathering, like the capping rock on the mesa (below left), while joints in the Hawkesbury sandstone at the Three Sisters (below right) let in the forces of weathering.

Two landforms shaped by weathering: a mesa in Nevada, USA (left), and the Three Sisters, Blue Mountains, NSW, Australia.

Anybody who looks after stone buildings knows a bit about weathering, and understands that it is a slow process. For example, the exposed parapet copings of Portland stone high up on the outside of St Paul's Cathedral in London have lost only 13 mm ("half an inch") in 250 years.

Do minerals weather? Surprisingly, they do, even the tough ones. The tagline "A diamond is forever…" was coined by Frances Gerety, for a de Beers' advertisements in 1947, but diamonds are just hard crystals, and way back in 1772, chemist Antoine Lavoisier showed that a diamond would burn, if he heated it enough. (In case you are rich enough to try this, Lavoisier sealed his diamond in a glass container and used a strong lens to focus the sun's heat on his target.)

Weathered limestone rock, Margaret River seashore, Western Australia. Without a scale, this resembles the volcano remnant at Cape Palliser (Agreement 7 in the introduction).

In geology, nothing is ever completely permanent. For starters, there is no such thing as "insoluble". Many minerals in rocks resist being dissolved, but over time, enough time, no mineral is ever totally insoluble. Some minerals are rather more soluble, and if any single mineral in a rock breaks down and washes out, it will only be a matter of time before even the hardest rock begins to crumble.

Air, heat and cold also play a major part in weathering. Geologists recognise two types of weathering: the visible one is physical weathering, sometimes called mechanical weathering, and this name probably tells us all we need to know about how it works. First, though, the rock usually needs to become weaker, which happens through chemical changes in the minerals, and that requires a few paragraphs of technicalities about chemical weathering.

Chemical weathering.

Most chemical weathering involves water, and chemical reactions go faster at higher temperatures, so the fastest breakdowns happen in warm and humid tropical climates. The simplest water weathering happens in those rare cases when there is a soluble salt in a rock, and this is dissolved out and carried away, as when halite, salt crystals are dissolved.

A pool in sandstone, Umina, north of Sydney. Water makes the damp sandstone weather faster.

Usually any soluble salts have long since gone, but if a piece of sandstone has its grains bound by calcium carbonate or limestone, the limestone "dissolves" quite fast. Note that this does not involve the rock dissolving in the way that salt or sugar will dissolve in water: it is a chemical change, and for those with some old-fashioned school chemistry, it can be written out like this:

$$CaCO_3 + H_2CO_3 \rightarrow Ca^{2+} + 2HCO_3^-$$

As a general rule, whenever you see a chemical group with a charge tacked on at the end, this means we are talking about charged atoms (or groups of atoms) called ions, which can slosh around in solution.

In chemical terns, ordinary limestone is fairly insoluble calcium carbonate, but even a dilute solution of carbon dioxide in water can react with limestone, turning the insoluble carbonate grouping into what used to be called the bicarbonate ion, though now we call it the hydrogen carbonate ion.

This ion (HCO_3^-) is less strongly bound to the calcium ion, and the two of them spread out into the water. If the water they are in evaporates, then calcium carbonate forms again, and we call the products stalactites and stalagmites—but we will come to those later.

Most chemical weathering results from hydrolysis, which means "splitting by water". A typical example of this is the breakdown of orthoclase or alkali feldspar ($KAlSi_3O_8$, which is found in igneous rocks) into a clay mineral, kaolinite ($H_4Al_2Si_2O_9$).

Oxygen is also important in some sorts of chemical weathering. When olivine, a common rock which is iron silicate Fe_2SiO_4 breaks down to limonite, $Fe_2O_3.H_2O$, oxidation has taken place. That leads us back to another water-involved chemical weathering, when limonite undergoes *dehydration* and forms brick-red hematite, which is just Fe_2O_3.

Because most chemical weathering involves oxygen or water getting into the rock, physical weathering opens entry points and so is very important.

Now we are done with the technical jargon for a while.

Physical weathering.

When desert sand is blown against a cliff and slowly knocks fragments off, that is physical weathering, but so is the action of water that soaks into the surface of a rock, high on a mountain, before the water freezes, expands, and wedges a few grains or pieces away.

As a rough guide, physical weathering involves one or more of heat, force, water or ice. The distinction between chemical and physical weathering isn't all that important, because all the different sorts of weathering go on at the same time. To be precise, chemical weathering involves rocks either losing existing minerals or acquiring new minerals, while physical weathering just involves the rocks being broken into smaller bits.

In summer, rock may be exposed to strong sunlight that makes the rock surface expand and chip away as well. Rocks are not very good conductors of heat, so under the surface, the rock is cool and keeps its former size, even as the surface expands in the sun. Stresses are set up, and once again, grains and fragments may fall off. Then there are stronger forces in play.

Heat can also come from lightning strikes, and around the world, there are about 100 lightning strikes, somewhere, each second. That adds up to a lot of energy hitting things. When lightning fails to hit a building, a tree, a foolish kite flyer or an unwise golfer, it usually hits rock.

Lightning often comes with rain, and when water has already soaked into a rock, the instant heat of a lightning strike turns that water to steam, flaking off a surface layer. With the rock in small pieces, other weathering effects take over.

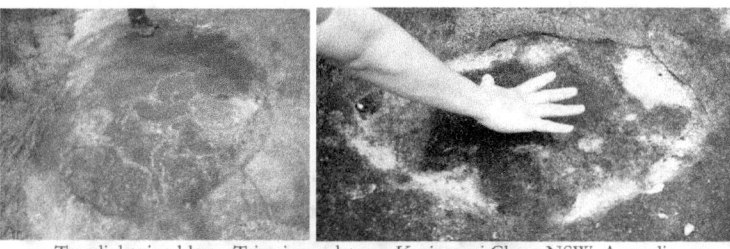

Two lightning blasts, Triassic sandstone, Kuring-gai Chase NSW, Australia.

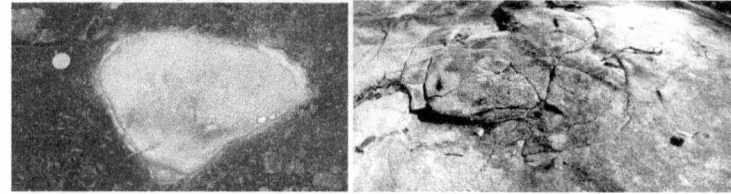

Two more lightning blasts, Permian sandstone, Budawang Ranges, Australia.

Joints and weathering.

Most rocks have planes of weakness through them called joints. Water can enter the rocks through joints and play a major part in the later weathering of those rocks.

Hawkesbury sandstone near Manly Beach, Sydney, Australia. Water in the joints has weathered the rock, and wave action has then carried the fragments away.

When geologists explain why joints form, they wave their hands quite a bit, and any keen student of scientific behaviour should know that when scientists wave their hands, it usually means they are unsure, but a probable cause of jointing is that the rocks are reacting to reduced pressure as the rocks above are carried away by weathering and erosion.

So let us just agree that jointing happens. We can understand how the cooling of a basalt flow produces columnar jointing as the hot rock contracts, but joints in sedimentary rock are harder to explain. There are often two sets of joints, more or less at right angles to each other, but the key phrase is *more or less*.

Another geological effect that provokes hand-waving comes when you ask about the wave-cut platforms that are common on many coasts.

A wave-cut rock platform near Ulladulla, NSW, Australia.

It is easy enough to see that the waves come rolling in over the platform to undercut the cliff behind, bringing down blocks of stone that are then beaten, battered and pulverised, but why is rock in the tidal range immune from weathering?

At high tide, waves come over the platform, filling the joints with water, and even carving the occasional rock pool where small animals shelter when the tide drops, but the rock mostly endures.

We can see a combination of weathering along the joints in Victoria in the area known (for tourism purposes) as the Twelve Apostles, though there have always been less than 12. The coast is limestone, and waves drive into the joints which run from the NE to the SW. Because waves refract around a headland, as the NW-SE cracks open, the waves swing around and cut in from the side, making caves which become an arch which later collapses to leave a stack.

Aerial shots, the Twelve Apostles, Victoria, Australia.

At the time when the photos above were taken, there were nine "apostles", but one of them collapsed in 2005. Don't worry, say the geologists: there will be more of them along—in (geological) time.

How granite tors form.

Remember the magical shapes of the granite tors that we skipped past earlier? Geologists won't call tors 'spherical' or 'round', but they will grudgingly concede that they are 'spheroidal'. That's near enough to round or spherical for us!

Granite tors, rural Victoria, Australia.

Granite boulders often have a spheroidal shape which gives the impression that somebody started off with cubes of rock and began chipping away at the corners. Well, as it happens, that's close to accurate.

The cubes come about because granite is a jointed rock, and while a piece of granite is still below the surface, water seeps in, and chemical weathering begins. Because the corners of a cube can be attacked from three directions at once, the corners weather fastest. When erosion brings the new blocks up to the surface, they are "born rounded".

The process may continue later, if the area is chilly enough to produce frost at times. If it gets cold enough, water that has soaked into the surface changes to ice, and may wedge particles of rock off. Any pointed bits and corners get more water, and are more exposed to the cold, so off they go!

Weathered granite, Freycinet Peninsula, Tasmania.

Wave Rock, Uluru and geological reasoning.

Even on a large scale, similar effects can be seen, but granite has more surprises to offer. Near Wave Rock in Western Australia, you can see spheroidal granite—and a truly amazing shape in the rocks.

Two views of the Wave Rock area, Western Australia.

These sights are some 350 km south-east of Perth, near the town of Hyden, and once you see Wave Rock, you will understand its name. Standing 11 to 12 metres high, the flared slope looks just like a giant wave, about to break. Most of Australia first discovered this rock in the 1960s, just after a bitumen road reached Hyden from Perth, making Wave Rock a feasible tourist attraction.

It was featured on 28 April 1965 in *Women's Weekly*, and the magazine said it was a result of wind erosion, though adding the suggestion that the shape might also be due to the 'action of glaciers'. Later popular stories said the cause was the action of an ancient sea, lapping the foot of the cliff.

Geologists pointed out that if the sea had been there, then there ought to be marine deposits around the place. These same spoilsports went on to explain that one feature of bare rock surfaces like Hyden Rock or Uluru was that a lot of water ran off them, when rare rains came by.

The soil around the rock got wet and supported more vegetation. So even if desert winds came rushing in, the plants would absorb the force of the wind, and stop the sand grains that were supposed to have shaped the 'wave'. Flared slopes are also found around other large bodies of rock, like the cave seen below, one of the ones around the base of Uluru that visitors are allowed to enter. Similar shapes can also be found on the Eyre Peninsula in South Australia.

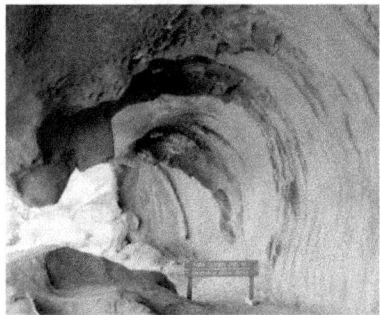

A weathering pattern at Uluru, Northern Territory, Australia, which reminds us of Wave Rock.

The simple answer is that the same water run-off that knocked out the sand-blasting theory was in fact a central part of the true cause of these concave forms. Australia is an old land, with old soils that have been exposed to dry winds over many years. Features like Wave Rock and Uluru that now rear up out of the flat plains were once hidden underground, within the plains. Over time, the winds uncovered them, but once the stone was exposed, rainfall ran straight off it.

The cleverest early white explorers quickly got the idea of asking the people who lived there, the Indigenous Australians, where to look for water. Some of them followed "native roads", knowing the tracks made by countless feet *had* to lead to water. Invariably, the "native wells", the soaks, lay at the foot of a rocky slope, and this steady supply of water explains how the wave shape developed.

Water weathers some of the minerals in rocks, whether the rock is the granite of Wave Rock, or the arkose sandstone of Uluru. The minerals break down, and in a sense, the rock just rots away under the influence of the damp soil. The minerals that had been rock became dust, and as the wind blew, or a rare flash flood gurgled across the plain, those minerals were carried away.

Other parts of Uluru (Northern Territory, Australia) show interesting weathering patterns.

Falling rocks.

One way of breaking up rocks is to drop other rocks on them. From time to time, rocks will fall from a cliff, or roll down a hill. This can happen after an earth tremor, or simply because the rock lying under it gave way, under the ravages of weathering. When this happens, some of the downhill rocks are about to take a pounding.

Fallen rock, Kata Tjuta area.

Philosophers like to argue about trees falling in a forest where nobody hears them fall. After a Wednesday in August 2016, they had a new conundrum, this one involving rocks falling and nobody hearing it. The location was my favourite playground, Sydney's North Head, a lump of Triassic sandstone with significant jointing. Joints will be explained in more detail in chapter 5, but for now, just think of them as vertical cracks.

Looking west, a view of the rock fall on the day of the fall.

At some point, some rock came down off the cliff, between the Hole in the Wall track and Fairfax Lookout. Perhaps somebody heard a bang, or two bangs, but that was it. Nobody seems to be sure about anything, and I don't report rumours, even though I react to them.

I picked up a rumour on the web, and hurried off to gather photographs. I was just in time, because the panic-merchants were already reacting wildly, fearing that Armageddon was upon us, we were all doomed, all of those things that make flailing mis-managers shout "don't panic!" to ensure that everybody else goes into a panic. (This is a cunning ploy to hide the fact that they started the panic!)

Temporary barriers were swiftly (well, three of four days later) put in place.

Several years on, the best access points are still blocked off. The shots above came from those two points, because I beat the authorities to it, assessed the safety, and went in to record an unusual event.

Closer views of the fallen rock.

The panic was based on the fear that "the whole cliff might come down". It will, one day, but not right now, and they blocked off unrelated bits of coast in any case. I gave up a flourishing career as a management consultant in 1990 to avoid dealing with mindless knee-jerk "managers" like these. To manage risks, you need to understand the facts and the principles. Here's what a typical cliff looks like:

At the bottom, wave action, and half-way up, differential weathering have undercut blocks of stone.

All rocks have joints in them, they are there, and rock can fall off when a joint is sufficiently undermined. Hawkesbury sandstone usually has two sets of joints, more or less at right angles to each other. Some of the sandstone beds are less resistant to weathering, and some of the beds in the sandstone are more like shale, eroding out even faster, undercutting the beds above. The joints shape Sydney's cliffs, keeping them vertical.

Inner North Head has two clear undercuts, as you can see in the composite shot above. When the undercutting goes right under a joint, the situation is ripe for a block to fall, and that is what happened.

I considered that it wasn't the whole cliff, just a block weighing perhaps 600 tons, a wildly inaccurate guesstimate, as it turned out, but this would still not have been nice to have land on you. My neighbour Geoff Lambert suspected that it was bigger. He did the research, using aerial photos, and came up with this:

> It was much bigger than I imagined. The surface area of the rock that fell was about 950m^2 and the height (if no overhang), was an average of 33m. Thus a volume of 31000m^3 and, at an assumed specific gravity of 2.5, a mass of about 75,000 tonnes.

That estimate was impressive, but still not a record. The last time Australia saw a fall of this sort was in January 1931, and it was called a landslide. The process was slower and better observed, beginning with a fissure or cleft near Dog Face Rock in the Blue Mountains, west of Sydney.

This opening went from 2 metres to 4.5 metres over a couple of days, and already, "hundreds of tons" had fallen by 27 January. Within 24 hours, an alleged 100,000 tonnes had fallen.

Sir Edgeworth David knew what was what: this process had shaped the valleys of the Blue Mountains, and it had been going on for millions of years, he told the *Sydney Morning Herald* on 4 May 1931. (See page 9 for the full story.

These events are rare, but inevitable, and for the past few years, I have been photographing likely future fall areas, in the hope of getting a before-and-after. In geological time scales, they are frequent, but on our scale, such falls are rare. The sky is not falling, Chicken Little!

At North Head, churning waves grind up the smaller sandstone fragments: note the colour in the water.

More commonly, rock falls in smaller fragments, tumbling down onto a slope that is called a talus slope in America, and a scree slope in nations where English is spoken. Under either name, these are usually not good places to walk, because the slopes are unstable. Some slopes are stabilised by the vegetation growing on them, but they are still dangerous.

Scree slopes near Sun Peaks, British Columbia and southern Croatia.

Another good place not to be is under a meteorite, and as they fall, meteorites also break rocks. Kaali järv ('Lake Kaali') on the Estonian island of Saaremaa shows what a meteorite can do.

Kaali järv, Saaremaa, Estonia.

Kaali is the largest of nine craters, and it is a circular pit, 110 metres across and some 22 metres deep: the other craters are all within a kilometre of this lake. Dating the impact is a challenge: the pit was *probably* created some 6000 ± 1000 years ago, though some people think it is only a little over 2000 years old.

What we know for certain is that a large lump came in at 10 to 20 km/second and broke up 5 or 10 kilometres up. The estimated mass is somewhere between 20 and 80 tonnes, delivering a punch close to that of the atomic bomb exploded over Hiroshima.

Two views of the Henbury meteorite craters, Northern Territory, Australia.

The Henbury craters are in a conservation area some 150 km south of Alice Springs, and the age estimates are all something more than 4000 years, but the event lives on in Indigenous oral tradition, and like the Estonian meteorite, the Henbury item broke up to form a dozen or so pieces of nickel-iron. About the only scientific information is that the temperature at the site went up to 850°C.

Honeycomb weathering.

Geology sculpts the rocks around us, and one of the delights of my home area is honeycomb weathering, sometimes called alveolar weathering, while others call it fretting, stone lattice, or most poetically, stone lace.

Honeycomb weathering in Hawkesbury sandstone, some of it cross-bedded, near Box Head, north of Sydney.

Much of the best-exposed sandstone near Sydney is close to the coast, and around the world, it is common to blame salt spray for honeycomb weathering. The idea is that salt spray lands on and soaks into the stone, then when the water evaporates, salt crystals are supposed to wedge sand grains off.

Honeycomb weathering on a coastal boulder, Maroubra, Sydney, Australia.

There is definitely more to the picture than that. While salt spray may play a part, as a young man, I saw honeycomb weathering in the Budawang Ranges, 40 km from the nearest sea coast. I have no surviving photographs from that time, but I do have something rather similar from the flanks of Uluru.

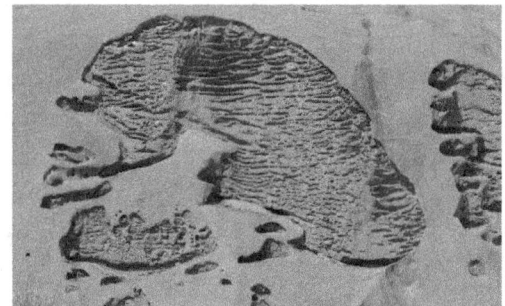
A curious but entirely natural weathering effect on the side of Uluru.

Honeycomb weathering, far from the sea: Cyprus (left) and Zion National Park, Utah (right).

Other theories put forward involve wind-scouring, or salts from the soil getting into the stone. In colder climates, frost wedging is sometimes invoked, with a tell-tale waving of the hands. My take on it: this stone is beautiful: why not sit back and enjoy it? Away from the coast, the best honeycomb weathering is often found in overhanging cliffs and caves.

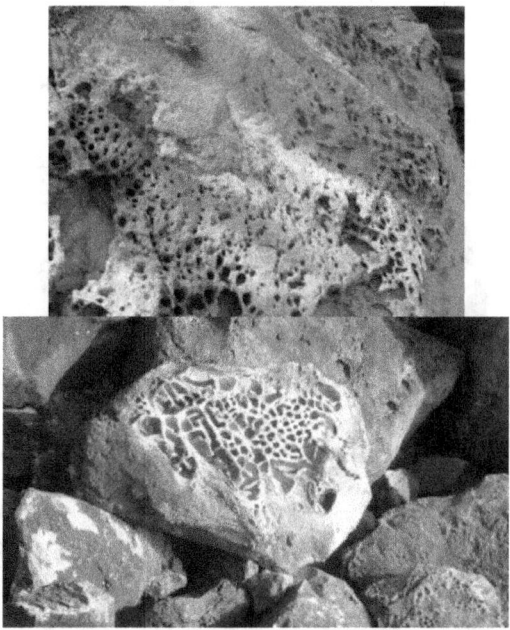

Coastal honeycomb weathering, Wobby Beach (left) and North Head (right).

If you are photographing honeycomb weathering, don't neglect to fiddle with the exposure, as reflected in the next two shots:

Two views of the same piece of rock, Malabar, south of Sydney.

Honeycomb weathering, Cuttagee Lagoon, NSW (left) and Kuring-gai Chase National Park, Sydney (right).

Hoodoos and pinnacles.

Differential weathering produces some true wonders, but before I turn to the best examples, I want to start with a small-scale example that is commonly seen on bush tracks where there is an exposure of clay soil that contains pebbles, at the side of the track. For scale in the picture below, the leaves in the foreground are about finger-sized.

Mini-pinnacles in clay soil. Each tiny pillar of soil is protected by a small pebble.

When it rains heavily, clay is splashed around, but when the wash-away level gets down to a pebble, the clay beneath is protected, even while the clay to either side is being washed away. The result is that tiny, fragile clay pinnacles begin to develop. In the state of Utah, one of the great tourist attractions is Bryce Canyon, which has truly amazing pinnacles.

Pinnacles ("hoodoos"), Bryce Canyon, Utah.

More hoodoos, Bryce Canyon, Utah.

Why are these pinnacles known locally as "hoodoos"? I don't know: they just are, but isn't it marvellous what a bit of weathering can do? I was told that most of the shaping was brought about by ice-wedging, as the area is cold enough for frost during a lot of the year.

Hoodoos in Red Canyon, Utah, and along a river bank in British Columbia.

Liesegang weathering.

Over the millennia, water seeps in through joints, and soaks into the rock. Given the right conditions, it may produce patterns like the ones below, which are called Liesegang patterns or rings, after the first scientist to study them, R. E. Liesegang, though his rings were generated by adding a crystal of silver nitrate (or a strong solution of silver nitrate) to a gel containing dilute potassium dichromate, which yielded periodic patterns of silver chromate as a precipitate.

Liesegang patterns, Swansea Heads, NSW.

This is another of those hand waving things, but my preferred explanation involves organic matter in sandstone that reduces fairly insoluble iron (Fe^{3+}) to soluble iron (Fe^{2+}), which then diffuses through the stone under the influence of water in the rock. At some point the Fe^{2+} is oxidised back to Fe^{3+}.

Well, just because I prefer it has no real bearing on the truth of that assertion, because it doesn't really explain the formation of the rings. The majority of the shots provided below were taken at the back of the rock platform near Reid's reserve in part of the Reid's Mistake Formation near Swansea Heads.

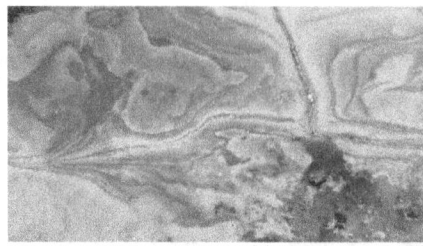 

Liesegang patterns, Swansea Heads, NSW.

This is a tuffaceous sandstone with pronounced jointing, and as I will describe in chapter 10, there are petrified trees there as well: in fact I was there looking for the trees when I chanced upon this small portion of the rocks. I have no more information to add: just enjoy the patterns!

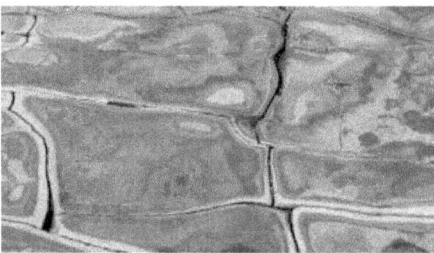 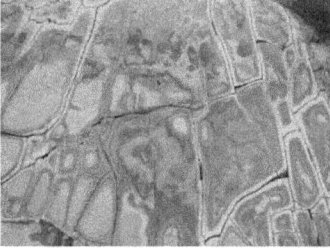

Liesegang patterns, Swansea Heads, NSW.

Now here are a few from another site:

Liesegang patterns, Malabar, NSW.

Glaciers and weathering.

In an ice age, up to a third of the earth may be covered by ice, flowing slowly across the land, pushing everything before it. Even today, 10% of the world's surface is covered in slow-moving ice, and 75% of our fresh water supplies are in glaciers which can be as much as 4200 metres thick in Antarctica, where some of the land appears to have been pushed down to 2.5 km below sea level, simply due to the mass of ice bearing down on it.

You might assume that the rock below the ice is protected from weathering, but glaciers don't just lie there. They flow under pressure, moving fastest at the top, midway between the sides, where the ice is furthest from any frictional effects. The base of a glacier is highly efficient as a grinder of the valley floor, with the rocks they drag over the lower surface, producing 'rock flour'. This is physical weathering, but the finely divided rock particles are more exposed than the bedrock they came from, which promotes chemical weathering. Past ice ages are detectable today because glacial erosion creates very distinctive landforms, indicating past history.

It took a while for humans to work out what had happened in the past. This is understandable, because while the Columbia glacier of Alaska moves 25 metres a day, most of the world's glaciers, especially those within easy reach of inhabited areas, are slower. Unless people looked very carefully, it would be hard to detect any movement in a glacier at all.

Glacial landscape, Glacier Bay, Alaska.

As a Swiss, Louis Agassiz saw plenty of glaciers, and in 1839, he learned that a cabin, built on a glacier in 1827, had moved about 1.5 kilometres in 12 years. He became curious about glaciers and established a camp, using a huge boulder and a blanket to form a summer shelter. He had hoped to use an old hut that he had seen the previous year, but he found the remains of the hut, crushed and scattered, some 60 metres down the glacier from where it had last been seen. At this point, he probably went *aha* (in one of the Swiss languages, no doubt!).

Towards the end of the season, Agassiz drilled a series of holes in a straight line across the surface of the glacier, then he carried up a set of stakes which he inserted into the holes, but there is no record of what he saw of them later, so they must have gone missing.

In September 1841, he drove a new set of stakes into the ice, which showed the next spring that the centre of a glacier flowed fastest. He added even more stakes in the summer of 1842, and so measured daily and nightly speeds and more. His proof that glaciers moved and a measure of how fast they moved, made his ideas spread at much faster than glacial speed.

A tidewater glacier drops into the sea, Glacier Bay, Alaska. The dark streaks are dust blown onto the glacier.

By now, Agassiz knew the tell-tale signs of glaciation, like a moraine, where a glacier had dropped the stones it carried, the U-shaped valleys and such, but very few scientists accepted his theory of a world shaped by ice ages, so he visited Scotland, and sought examples there, though he could also have gone to Canada.

Only a few of those he spoke to in Scotland accepted his Ice Age notions, but those who did were influential, and his ideas were slowly accepted. These days, everybody knows about glaciers and ice ages. Some of the world's most approachable glaciers are the tidewater glaciers, where you can see the face from a boat or ship.

The Margerie Glacier, a 34 km long tidewater glacier.

One reason why people worry about global warming is that the world's ice, if it all melted, would make the sea levels of the world rise by about 70 metres, but contrary to public assumptions, the tidewater glaciers play only a small part in the rising sea levels: the big problem lies in the melting and retreating glaciers in higher areas, up in the mountains.

Outside of Antarctica and the Greenland Ice Sheet, global warming from the greenhouse effect continues and the volume of the world's glaciers outside of these areas continues to decline. The smaller, high-altitude but low-latitude glaciers seem to be taking the biggest hit.

Africa's Mt Kenya lost 92% of its glacier mass in the 20th century, and Mount Kilimanjaro glaciers have shrunk by 73% in the same time. There were 27 glaciers in Spain in 1980, but now that number has since dropped to just 13.

The European Alps have undergone an ice loss of about 50% in the past century, and New Zealand glaciers have shrunk about 26% since 1890. Only a few hundred of the world's estimated 200,000 glaciers have been studied in detail, but the pattern is uniform.

The world's glaciers outside of Antarctica and Greenland make up only about 6 percent of the world's total ice mass, but the water from these is recycled more quickly and contributes more to sea level rise than do the polar ice sheets, so any ice loss needs to be a matter for serious concern.

A lot of the world's best agricultural land, many of its cities and all of its ports would be flooded by a rise of even a metre or two. In the Pacific, several whole nations would be flooded as their land surface, made of coral atolls, is generally just metres above the high water mark.

The Thames Barrier in London, installed in 1986 to delay large tides and storm surges.

Parts of the London Tube (Underground rail) will be flooded, coastal roads and wharves will be closed, airports will be under water: just look around you, to see what is at risk. The only people left opposing this assessment are the terminally stupid, the corrupt and bought commentators and the obscenely greedy. If you don't like this conclusion, read chapter 16, and if you still don't like it, get your stupid grubby nose out of my book!

The most serious aspect is the certainty of future wars about water. With glaciers melting, all over the world, there will be murderous summer droughts in areas like India which rely on the release of summer meltwater. The coming refugee problems will be worse than what we have seen so far in the 21st century.

The main biological significance of ice age glaciers lies in the way they grind off the surface of the earth over large areas, removing all soil, and all traces of life. At the same time, each glacier exposes fresh rock, from which fresh soil can develop when the glaciers retreat: Australia, with very few glaciers, has many areas where the soil is old, tired, and lacking in essential minerals.

Glaciers are found in many parts of the world, even close to the equator in Irian Jaya (West Papua). Other well-known glaciers include the Zermatt, Stechelberg, Grindelwald, Trient, Les Diablerets, and Rhône in Switzerland; the Bossons Glacier in France; the Nigards, Gaupne, Fanarak, Lom, and Bover in Norway; the Tasman glacier of New Zealand, the Wright, Taylor, and Wilson Piedmont glaciers in Antarctica; and in north America, the Emmons and Nisqually glaciers on Mt. Rainier, Washington; Grinnell glacier in Glacier National Park, Montana; the Dinwoody glacier in the Wind River Mountains and the Teton glacier in Teton National Park, both in Wyoming.

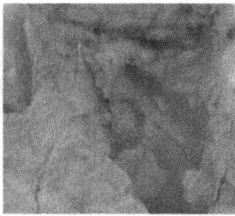

Iceberg with blue patches, Glacier Bay, Alaska. The close-up shows the blue is not an optical illusion.

Now a note which will only make sense when I turn later to how geologists find things out. It is relevant there, but it belongs here. Glaciers often appear blue, because they reflect blue light, and absorb all other colours, but they are really made of nothing more than ice.

Most glaciers travel slowly, although other "surging" glaciers like Pakistan's Kutiah glacier exist, and in some of them, the ice has advanced as much as five metres in an hour, but the mechanism behind this is largely unknown. Some scientists believe (or fear) that the West Antarctic ice sheet may be able to surge, delivering huge volumes of ice to the world's oceans in a period as small as 100 years, causing significant climate change as it did so.

The movement of the ice over an uneven base can cause cracking, compression and extension of the ice, an effect which is most common at the head of a glacier, and at the snout of the glacier, where ice loss balances the glacier's forward progress, an area called the ablation zone.

The ablation zone of Briksdal glacier, Norway.

The glacial ice can slide over the bedrock, especially where the pressure melting point is reached, so the movement is lubricated by a film of water. Most glaciers move by a combination of these different effects. Glaciers flow under the extreme pressure of the ice on top, usually slowly, but the Kutiah glacier went 12 kilometres in three months during 1953, or more than 100 metres each day.

Glacial landforms.

The most obvious glacial landforms are moraines, cirques, drumlins, hanging valleys and *'erratics'*, which remain long after the glaciers are gone. One of the best indications of past glaciation comes from wide U-shaped valleys, quite unlike the V-shaped valleys made by flowing water in streams and rivers.

A set of hanging U-shaped glacial valleys near Akureyri, Iceland.

A glacial valley in the High Andes. Note the erratics in the foreground, stones scattered across the valley floor.

When a valley glacier stops dragging material along for any reason, a build-up of broken rock occurs, forming a wall-like structure called a moraine. These are named according to their position in relation to the glacier.

(Left) Angel Glacier and moraine, Mt Edith Cavell, British Columbia, and (right) its valley.

A **terminal moraine** forms where most of the material is dumped at the front of the glacier. This forms when the glacier is melting as fast as it moves forward. This moraine debris may be pushed forward in winter when the glacier advances slowly, and it may be scattered in front of the main wall in summer as the glacier retreats, only to be bulldozed forward again in the following winter, or summer may see a rush of meltwater which carries the moraine materials away to be deposited as till.

After the glacier has finally retreated or disappeared, the moraine will typically be cut through at one point by the stream or river which replaces the glacier in the valley, leaving a steep-sided valley passing through a wall of rubble blocking the broader U-shaped valley left by the glacier.

Moraine Lake, Canada, and glacial traces in the high Andes, Peru.

Varves are the geological equivalent of tree rings. A varve is a layer of sediment deposited in a lake during one year. Each layer has two parts, deposited at different seasons and they differ in in colour and texture, so the layers can be counted and measured.

A block of varved shale with a normal fault, Sandy Hollow, New South Wales.

This results in a series of annual layers which can be used like the growth rings of a tree to estimate past climate, since varves are typically laid down either in areas where the water freezes in winter, or glacial areas, and in areas where rainfall is distinctly seasonal.

The layers form varved shales, which may be used to assess the former climate of an area, allowing for the fact that the match-up between water flow from a glacier and temperature can be partly buried in random variables such as rainfall patterns.

Fjords are particular features of three coastlines, Norway, the south island of New Zealand, and East Greenland. A fjord (or fiord) is a long, narrow, steep-sided coastal inlet extending far inland and often very deep. Most fjords are drowned valleys formed by glacial erosion, and filled with sea water as the sea levels rose again, after the glaciers retreated.

Glaciers can carve out rock well below sea level, because the ice is held down on the valley floor by the weight of ice above, cutting it deeper and deeper. As the fjord glaciers warm and lose surface ice, they begin to "calve", to drop off smaller icebergs, retreating rapidly as the sea pushes in under the ice, lifting it and breaking it away.

Icebergs.

Jökulsárlón Lagoon in Iceland and it icebergs, 'bergy bits' and 'growlers'.

Icebergs are pieces of ice, more than 5 metres across, usually found at high latitudes, either close to Antarctica or in the North Atlantic, and they break off from glaciers or ice caps. Icebergs can be hard to find and study, but if you go to Jökulsárlón Lagoon in Iceland, it's almost like shooting fish in a barrel. The Breiðamerkurjökull Glacier is part of the Vatnajökull Glacier, and the lagoon's blue waters are covered in icebergs, while the shores are dotted with tourists.

One of the strangest things about ice is that it floats on water. Usually, as things get colder, they contract, getting smaller and denser, so we would usually expect solid ice to sink, but water is different. It contracts and gets more dense until it reaches 4°C, and then it expands again, and it expands remarkably as the water molecules lock into a crystal lattice with spaces, which is why ice floats.

With only a fraction of their mass above water level, icebergs from Greenland are a danger to shipping, particularly in the North Atlantic. Antarctic icebergs are characteristically huge and flat. Some of the Antarctic ones can be 80 km long, but even the piece of the iceberg that we see is only a small part of the whole. As the old saying implies, we only ever see the tip of the iceberg, about 10% of it. Ice is less dense than water, and sea water is more dense than the fresh water which makes up an iceberg.

Icebergs carry samples of rock and other sediment. As the originating glaciers travel over the ground, they collect sediment, which remains stuck to the underside of an iceberg for some time, often allowing the origin of an iceberg to be determined. Remember this when we look at a dropstone in chapter 10.

The name of an Antarctic iceberg is derived from the quadrant it is found in:

> A = 0° to 90°W (Bellinghausen/Weddell Sea);
> B = 90°W to 180° (Amundsen/Eastern Ross Sea);
> C = 180° to 90°E (Western Ross Sea/Wilkes Land);
> D = 90°E to 0° (Amery/Eastern Weddell Sea).

The number which follows the letter is issued in a simple sequence. If A-38 breaks up, the pieces will then become A-38A, A-38B and so on. The number of icebergs seen outside the Arctic and Antarctic Circles has increased with global warming, because this causes glaciers to lose ice faster.

On occasions, iceberg speeds in the ocean can get above 1 m/s, 3.6 km/h, but even that adds up. In the course of a day, an iceberg has the potential to move almost a degree of latitude and well over one degree of longitude in waters close to the polar regions where the meridians are close together.

In the Antarctic, if an iceberg is caught up in the Antarctic Circumpolar Current, it may take off at an astounding 7 km/h, making tracking quite a challenge. Any penguin hitching a ride on such a lump of ice should swim back, rather than go with the floe.

Anatomy of an ice age.

There is enough evidence around to show that there have been many past ice ages, even one in Australia, in the Permian, about 270 million years ago. I have visited the site at Selwyn Rock near the Inman River on South Australia's

Fleurieu Peninsula, but my photographs are unconvincing. This Finnish one is better, but you still need to look hard.

Near the Sibelius memorial in Helsinki, Finland, glacial striations.

One thing is clear: there have been many times in the Earth's history when ice sheets and glaciers have advanced from polar regions to cover areas previously of temperate climate. The most recent (the one we call "*the* Ice Age") being from about 1 million years ago, up until around 10,000 years ago, when the ice retreated to its present polar position. Ice ages certainly happened as far back as the Permian, and may even have been extreme enough to have triggered something called Snowball Earth. We will look at that in chapter 15.

The ice ages are generally seen as a time of glaciers, mainly because this is the northern hemisphere experience. In the southern hemisphere, the ice ages saw much smaller glaciers, and the main effect was for the weather to be cooler, and much drier.

Nobody knows what causes an ice age to start, but one possibility is increased dust in the atmosphere, either from a volcano, or from a large meteorite impact. This dust would then increase the world's reflectivity or albedo, cooling the planet down, and causing more snow and ice to form, increasing the earth's albedo still further.

The main problem with this theory is that we do not know what would then make the ice age finish. Maybe dead organic material, trapped under the ice, breaks down to form methane, or ocean currents switch, warming and melting the ice, or maybe it is something completely different, like CO_2 outgassing from a volcano. On cooler continental shelves, there are reservoirs of methane clathrates, best described as methane ice. The lowering of sea levels may have caused these to break up, releasing methane into the atmosphere. It is worth noting here that greedy idiots in the USA have previously proposed drilling into these reservoirs, and under the present feckless US administration, this may bob up again. In 2017, Japan and China were also playing with this crazy idea.

Between the 17th century and the late 19th century, the world went through a 'Little Ice Age', when temperatures were cool enough for significant glacier advances. This may have been caused by lower solar activity, or any of a range of other astronomical effects. Popular candidates for possible causes include slight changes in the earth's orbit, wobbles and precessions in the planet's spin, and maybe even the earth passing through clouds of stellar dust.

The most extreme point in an ice age is referred to as the glacial maximum, the point at which the glaciers stop advancing. As soon as they begin to retreat and lose ice from the top, an isostatic uplift called glacial rebound operates. Scandinavia, for example, has less to fear than most parts of the world from global warming, as the area is still rising out of the sea. The Little Ice Age showed three maxima, beginning about 1650, about 1770, and 1850, each separated by slight warming intervals.

Evidence for past ice ages.

The principle of uniformitarianism applies in most things: we assume that the landforms shaped in the past were carved by the same forces we see around us today. When we find a till (a sediment or drift made up of an unstratified and unsorted deposit of clay, sand, gravel, and boulders), we assume that it has been left behind after the retreat of glaciers and ice-sheets, because we can see the same process happening today.

Till is easy to recognize because the till minerals, freshly ripped from the bedrock beneath the glacier, have been dragged or washed just a short distance from the glacier's snout before being dumped in a heap of unweathered bits and pieces. In the case of older glaciations, we may find tillite, which is just consolidated till, but with the same characteristics.

Glacial till in a road cutting near Quito, Ecuador.

And that brings us to soil, silt and sand.

4: Spicks and specks.

While sediments aren't rocks, you can't do much geology without the bits and pieces that come from rocks, and go back to rocks. Sediments drive the rock cycle, the basic process that underlies all of geology. Besides, sand is fun, if you have a hand lens or microscope.

Assorted beach sand samples, under the microscope.

Erosion.

> The United States alone, it is estimated by Federal geologists, is robbed of 783 million tons of its native soil every year in this way.
> —A. W. Haslett, *Unsolved Problems of Science*, London 1937.

Consider the desert landscapes below, one from Todra Gorge in Morocco, one from the Kimberleys of Western Australia, and two from western USA. Ask yourself: what shaped the land in these dry places?

Todra Gorge, Morocco, with a tiny stream through it. The Bungle Bungles, from the air, NW Western Australia.

The Grand Canyon (left) and desert country near Las Vegas.

Where there are seas or rivers running, it is easy to account for the movement of sediment, and in some cases, we can even see traces of the water that did it, but consider the two artificial examples that appear below. One is a mine's spoil heap, and the other is what was left after "hydraulicking", which involves washing sediment away to win the gold that is within it.

Spoil heap at Leigh Creek, South Australia (left) and remnants at the Oriental Claims gold field, Omeo in Victoria.

Human-driven erosion sends much more sediment into waterways than happens under normal conditions, though in the 19th century, people who engaged in sluicing and dredging for gold argued that the river was the natural place for sludge (mud) to go. Mind you, strychnine, funnelweb venom, cholera, smallpox, measles, lightning, earthquakes and tornadoes are all 'natural', as well.

In other cases, the stream that did the heavy work is clearly in view, as in the natural bridge in Hawkesbury sandstone below. There is enough room under the bridge for a short adult (like myself) to walk through without stooping.

Natural bridge, French's Forest, Sydney, Australia.

Coastal wave action shapes the headlands and the beaches.

Water may run fast through gorges and valleys, but wave power can be channelled, if the waves are directed into joints. Below, you see Australia's famous Kiama Blowhole, and also an entirely unknown small blowhole near my home. The famous one spouts up to 20 metres high, the other rarely reaches 20 cm, but the mechanism is the same: a wave thumps in, compresses air, and then some of the water is blasted up into the air.

Blowhole, Kiama, NSW, (left) and small blowhole, Delwood Beach, Sydney, Australia.

In some places, glaciers are responsible for moving the sediment, but at other times, wind is the main agent. In September 2009, Sydneysiders awoke to a very strange light, a red haze that reduced visibility. It was a dust storm, taking dust from somewhere in the southern Lake Eyre Basin in central Australia and possibly western and central New South Wales.

Sydney's 2009 dust storm.

The newspapers dubbed it "the Red Dawn". There had been winds in the 80 to 100 km/h range in western New South Wales, and the storm also affected much of the east coast of Australia. Later estimates suggested that the event dumped 2.5 million tonnes of sediment into the Pacific Ocean.

One curious effect of this event was a bloom of phytoplankton in the Tasman Sea as the iron-rich dust made extra iron available to them. No incident ever has just one result…

Two shots from the same position, during and after the famous Sydney dust storm.

Most people think the Sahara is nothing but sand, like the dunes seen in the next photograph, but the foreground shows pebbles. Any consumer of Boys' Own desert fiction of the P. C. Wren variety knows all about sand storms.

The author, searching for spiders in the Sahara, Morocco.

Indeed, the sixth paragraph of *Beau Geste* raises just this stereotype:

> And across all the *Harmattan* was blowing hard, that terrible wind that carries the Saharan dust a hundred miles to sea, not so much as a sand-storm, but as a mist or fog of dust as fine as flour, filling the eyes, the lungs, the pores of the skin, the nose and throat; getting into the locks of rifles, the works of watches and cameras, defiling water, food and everything else; rendering life a burden and a curse.
> —P. C. Wren, *Beau Geste*, chapter 1.

So in the Sahara, famously parched (though not always, as the picture below attests), and in the arid zones of Nevada and Australia, wind does a lot of the moving of sediment. When rain does come in the desert, it usually comes as a flash flood that moves the bigger rocks and boulders.

The author's wife's feet, following a storm near Aït Benhaddou, not that far from the Sahara in Morocco.

The particles that make rocks.

Sediments come in different sizes that have names, if you talk to geologists:

Boulders are greater than 256 mm diameter, bigger than your head;

Cobbles are particles between 64 mm and 256 mm in diameter, fist size to head size;

Pebbles are between 4 and 64 mm in diameter, the ones people like to throw into the water;

Granules are particles between 2 mm and 4 mm in diameter that will probably crunch under foot;

Beach sand can often be largely shell grit.

Sand has particles 0.0625–2.0 mm in diameter, big enough to see minerals;

Silt particles are between 0.0625 and 0.004 mm in diameter, too small to identify and tasting gritty;

Clay particles between 0.004 mm and 0.002 mm in diameter, and they stay in suspension;

Colloids are very fine sediment particles less than 0.002 mm in diameter, and they stay in suspension;

Mud particles also have a diameter of less than 0.002 mm, too small to see, and it tastes smooth, geologists say;

Gravel and **stones** are tricky: gravel is a mixture of particles between 256 mm and 2 mm in diameter, and stones are particles larger than 20 mm in diameter.

Pebbles, Aragunnu Beach, NSW south coast.

The particle size in a sedimentary rock depends on the speed of the wind or water flow, and coarser sediment settles as the speed drops away, producing graded deposits. Anything down to pebbles forms a rock called conglomerate, sand forms sandstone, and smaller particles form shale, mudstone or siltstone. If particles come from a volcano as 'ash', that's a tuff, or perhaps a breccia.

Now it is time to look at what happens to the sediment, once it is moved, starting with beaches.

The shapes of beaches.

Just before Christmas 2019, I was walking along the shore to a favourite restaurant for lunch, and we were discussing teaching possums to fly underwater, as one does. The tide was in, waves were crashing into the seawall, and this diverted my thoughts. "Why do waves always slope up to the back?" I asked.

Without missing a beat, my wife said "Maybe because they'd look silly, sloping *down* to the back?" At this point, a friend wandered by, so we asked her, and she thought it might be "wave action", which is probably right, but it's less than a complete answer.

So even though this chapter is about sediments, we need to look at waves as well. As it happens, I like throwing pebbles into calm water and watching the ripples spread, and we can learn a lot from that, mainly about the reasons beaches are curved, as the pictures below show:

The progression of ripples after throwing a pebble into calm water, Lake Visser in Norway. Use the background cloud to see the changing scale.

The main thing to notice is that the outside waves at the start are much higher than the later ones. The energy of the wave stays the same, but as the ripple's diameter gets larger, the energy is spread out over a much larger circle.

The next thing to notice is the circular shape, the last is that there are still new waves starting in the middle. They may be smaller than the first waves, but they are there. Try tossing a pebble in, and see how long the ripples last.

Ocean waves begin, far out to sea, as the wind makes circular ripples, but they soon combine with other circular ripples to make a straight wave front that comes pushing into the shore, where tricky effects make it curve again.

A scientist will tell you that waves are caused by water molecules going around in circles. The water of the wave never moves, but the average scientist can only prove this with a lot of mathematics, and a fair amount of hand-waving.

The simplest way to explain waves is to say they move energy from one place to another, and except near the shore, waves don't move any water around.

Waves change as they reach the shore. Only now does some of the water rush forward as the wave breaks.

A surf wave breaks when the water depth is less than one-seventh of the distance between wave crests. This tells us the waves interact with the sea bottom, and this is why beaches are curved.

When any wave hits the beach on an angle, it pushes sand sideways, making the beach change shape. The sand only stops moving along the beach when the beach is curved enough to always meet the waves square on. The beach shapes the waves, and the waves shape the beach until they settle into a stable pattern.

Looking down from Barrenjoey, north of Sydney, Palm Beach is on the left, Pittwater is on the right. Wave action shapes the beaches even in calm waters, like those of Pittwater.

Just about every beach you look at has that same curved shape. There are a few exceptions: if you look up <"90 mile beach">, you will find two long and very straight beaches, one in Australia and one in New Zealand, but even on those, at the ends, they still curl into a smiley shape, if you are looking out from the land side. Neither beach is 90 miles long, though!

And now, let's look at the rocks that sediments sometimes form.

5: Rocks in beds.

Ziz valley, Morocco.

Speaking formally, sedimentary rocks form when particles are transported and dropped by wind, water or glacial ice, or by precipitation from solution in water under normal surface temperatures and pressures, or by the aggregation of inorganic material from skeletal remains. Less formally, stuff is carried in, dropped, covered and buried. The stuff becomes sedimentary rock.

Sediments, compressed and heated gently, form sedimentary rocks, which can be weathered and eroded to form yet more sediments. Examples include shale, limestone and sandstone, and telling them apart sometimes calls for a bit of geochemistry, or a good look at a freshly broken surface.

The formal definition in the first paragraph rules coal out as a sedimentary rock, but it is often treated as sedimentary as well. That approach makes sense when you have coal beds lying between beds of undoubted sedimentary rock.

(Keep in mind that the handy labels we use are a bit artificial. When volcanic ash settles to make a layer that later becomes rock, is it sedimentary or igneous? Most of my geological friends say it is sedimenteous or ignentary. In short, the classification of coal is the least of your worries.)

In the 1700s, when rocks were first grouped together, they were looked at in terms of what ores they contained, or what they might be used for. Sandstone might be simply referred to as 'freestone', because it was usually free of the close laminar beds found in shales and some limestones, but you need to be careful:

> A *freestone* is a uniform thick-bedded sandstone with few divisional planes. It can be cut or worked easily in any direction, and consequently forms a good building stone. The term freestone is also applied to some limestones of similar character.
> —G. W. Tyrrell, *The Principles of Petrology*, Methuen, 1929, 210.

Stonemasons, of course, still distinguish one stone from another by their working qualities, but until rocks were classified as being of certain types, there was no chance of their origins being determined or scientifically studied.

When lime and sand are mixed, you may have a calcareous sandstone or an arenaceous limestone. The moral: nature does not work by pure categories, and it is unwise to get too hung up on the names, because the outline above is only the tip of the iceberg, and there *might* be a polar bear behind you on the berg...

Curiously shaped iceberg, Tracy Arm, Alaska. Look closely!

Now before we move on to the truly unusual structures, a quick word about some strata of the Grand Canyon in the USA, and some apparently normal rock from northern Australia, rock which conceals a secret.

Strata, Grand Canyon, USA (left) and the Bungle Bungles Australia (right).

These may look like two rather similar formations, but the Bungle Bungles in the north-east of Western Australia, have unusual orange and dark grey stripes. The rocks are late Devonian, and the dark layers have a higher clay content which holds on to moisture better, providing a good base for cyanobacteria, the simple ancient organisms that used to be called blue-green algae.

The surface layer of cyanobacteria is shallow, but enough to protect the rock and stop it eroding. The lighter coloured layers have a lower clay content, meaning they are more porous, so they dry out faster and provide a poor habitat for the cyanobacteria. The surface is more exposed, so iron in the rock oxidises and gives the rock its rusty-orange colour.

At the times when tourists go there, the land is often hot, dry and parched, but over the summer, the monsoon rains (called 'the wet season', or more colloquially, 'the wet') come roaring through, cascading down the rocks and washing out the chasms.

The beehive shapes of the Bungle Bungles have formed over the past 20 million years or so, by a combination of uplift, weathering and erosion, and the process is still going on today.

Sandstone.

This sandstone also contains quartz pebbles. It is mainly sand, hence its name.

Sandstone is a sedimentary rock made mainly of small grains of silicon dioxide (quartz) sand. It also occurs as two recognised variants, arkose and greywacke. Arkose is sandstone where the quartz is accompanied by feldspars. Greywacke is effectively dirty sandstone, where quartz, feldspars, rock fragments and volcanic debris have been thrown in.

Many sandstones are both porous and permeable, so they act as aquifers, meaning water-bearing rocks. We will come back to this in chapter 9.

It is usual to say that sandstones do not contain fossils, but in the 1940s, fossils were found in six hundred million year old rocks in the Ediacara Hills, outside Adelaide. There, in ordinary sandstone, were impressions of animals like modern jellyfish. These were fairly simple prints, and it was several years before people realised there were many other animal "prints" in the Ediacaran rocks.

Now, the Ediacara fossils are famous as the world's best collection of Pre-Cambrian remains. They are not, however, the only fossils, and it was the fossils of the Old Red Sandstone of Britain that began the fossil craze, and our sense of geological history, a century earlier.

The Old Red sandstone is remarkably rich in fish, and lies below the rocks that represent the age of reptiles. Mammals seemed to leave their fossils in younger rocks still. But below the Old Red Sandstone, the rocks seemed to contain no vertebrate fossils at all.

The Hawkesbury sandstone in my hometown of Sydney appears to have been laid down in a Triassic delta, rather like Bangladesh today, with a huge river braiding back and forth, washing out the finest minerals, the clay and other mineral-rich sediments, and leaving just the quartz grains behind.

The grains were rounded, and had probably been in an earlier sandstone somewhere else, but they settled where Sydney is now, almost 200 million years ago, waiting to play their part in shaping modern Sydney. Some of the sandstone beds are better bonded than the others within the Hawkesbury sandstone, but they are otherwise much the same, right through the deposit.

Back when the sediment beds were being deposited as huge sandbanks, one effect of the braiding, winding rivers is that sometimes a new stream bed was cut into the sand, and later filled in. These are called washouts.

A washout in Hawkesbury sandstone, in the Blue Mountains.

Please treat the next two pictures as illustrations or models, not the real thing, because neither of them shows sandstone in the making:

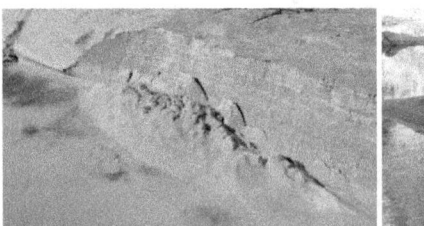

A sand deposit, The Entrance, NSW (left) and a sand bank in a creek, Cape York.

Sandstone does not usually come from the sand you see on beaches. Rather, it forms from sand which has washed down huge rivers, so the sorts of wind patterns you see below rarely show up in sandstone.

Sand derived from volcanic rock, Karekare Beach, New Zealand.

The evidence for the origins of a rock can come from many sources, but one of the most convincing comes from fossils that you find in the rock. The ripple marks found on the top of a sandstone slab can be even more convincing.

Charles Lyell's example of preserved ripple marks in New Red Sandstone from Cheshire in England (left), and a find from a creek bed near El Questro in the Kimberley Ranges of W.A. (right).

British geologist Henry Clifton Sorby started ripple research in 1859. He used to say later, after the worth of his work was recognised, that people laughed at him for trying "to examine mountains with microscopes", but his work was typical of the fine detail often required to tease out the truth in science.

Sand ripples in a sandbank, Karekare Beach, New Zealand.

Sorby examined the ripple patterns left in sedimentary rocks formed in shallow water. He assumed uniformity, that conditions when the rock was laid down were similar to those of today. From that base, he described how the rocks formed, and even reported the direction the waves and currents had travelled.

Two views of sand ripples, Santa Cruz, Galapagos Islands. How do these ripples compare with the New Zealand ones?

As it happens, you can look at cross bedding and discover how water flowed, back when the beds were laid down. One of my geology teachers earned his PhD for doing precisely that.

But what is cross bedding? Read on…

Cross bedding.

Sandstone often shows this form of bedding (sometimes called current bedding), with sloping beds sandwiched in between horizontal beds.

Triassic Hawkesbury sandstone, Old man's Hat, North Head, Sydney, Australia.

In 1669 Nikolas Steno, in his *Prodromus*, suggested that tilted strata in geology were originally laid down horizontally, and later lifted up by some force, but it needn't be so. When sediment is carried to the front of an advancing bank of sediment and pushed over the edge, it forms a characteristic slope at the angle of rest or repose.

Sandstone often shows what I prefer to call cross bedding, even though it is formed when sediment is pushed into place by air or water currents. It is seen where the horizontal layer that was a sand bank has internal beds laid down at the angle of rest, of around 30°. Look for horizontal beds, above and below.

Hawkesbury sandstone with cross bedding, Hickson Road, Sydney, Australia.

In the picture above, a water current moving from right to left once pushed new sand along until it tumbles down the front edge of the sand bank, making a sloping layer. The individual beds often have horizontal layers above and below, and each bed is usually about 50 to 60 cm thick.

Aeolian sandstones, rocks which are formed in wind-blown desert sand, can show the same sloping effect, but the beds may be 20 metres or more deep.

Cross-bedded aeolian sandstone, a fossil sand dune, Zion National Park, Utah, USA. Using the trees for scale, notice how deep this bed is.

Sand dunes and sand banks all show this characteristic angle of rest (or angle of repose). This is the steepest angle that loose dry particles can form, and it depends on local gravity, the attractive forces between the particles, the shape and size of the particles, and partly on the friction between the particles. These variables determine the angle of rest in a particular case.

Mojave desert dunes at dusk.

A slope at the angle of rest is unstable. If you dig a small hole at the base of a dune, or tread on it, a whole load of sand comes sliding down at you. People travelling in inland Australia on foot know that when you step onto a dune, the sand collapses beneath you, so climbing a 20-metre dune feels like climbing a hill three times as high, and it's the same all over the world.

Most undisturbed sand dunes lie at this angle, because sand is blown over the top of the dune. If the slope is less than the angle of rest, the sand grain stays at the top, if not, it rolls down the slope until it comes to rest. Wheat, gravel or sugar, all sorts of matter will develop their own unique maximum slope, and even scree slopes do the same. (This is what makes scree slopes dangerous.)

On the edge of the Sahara in Morocco, advancing sand lies at the angle of rest.

You can also see the angle of rest in ant lion pits, which can be found all over Australia: the pits are about the size of a 20-cent coin. You can explore the angle of rest in sand, rice, sugar grains or anything else with a simple cylindrical jar: the one in the picture has dry sand in it.

(Left) an ant lion pit: the sides are at the angle of rest; (right) my angle of rest measurer.

Now let's turn back to the aeolian sandstones of Utah. About a dozen years ago, a study of the zircons in these desert sandstones revealed that the sand in the stone came from the Appalachian Mountains in eastern North America.

The sand must have been carried in an ancient westward-flowing river system that transported sediment across the continent of North America. This was during the Jurassic era, approximately 190 million years ago, but the river was similar to the Amazon in modern day South America, which carries material the other way, east, across that continent from the Andes.

During the Jurassic, a huge sandy plain, similar to today's Sahara, covered much of the western United States. If the beds below look suspiciously as though they might have been laid down in water, they weren't. These beds below are aeolian (or *eolian*, if you are American).

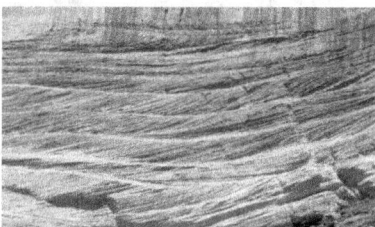

Cross-bedded Navajo sandstone, Zion National Park.

In southern Utah, these desert sands are exposed at Zion National Park in the Navajo Sandstone, a rock unit famous for its large fossil sand dunes visible in steep cliff exposures.

As Charles Lyell showed, tilted strata, whether they were originally horizontal or laid down as cross-beds, can be deceptive. From some angles, they may even appear horizontal. If you take a book and sit it on a table with the top end on a block of wood, obviously all the pages are sloping, but if you move around so you are looking into the top edge of the book, the page edges will appear horizontal. Lyell demonstrated the effect like this:

How Charles Lyell drew the world's attention to the role of point-of-view in geology.

I managed to capture this effect in real life, not far from my home, at a point where cross-bedded sandstone had been artificially cut at some point in the past. Notice the two contrasting dips on the two edges:

Two sections through a cross-bedded stratum in one shot, Sydney Road, Fairlight, Sydney, Australia. The shaded edge (to the left of the corner) appears to have almost horizontal beds.

Cross bedding is something of a major temporary obsession with me, so here are a few more examples:

Cross-bedded Hawkesbury sandstone, Malabar, Sydney (left) and on the right, another view of the Fairlight Sydney exposure, seen above.

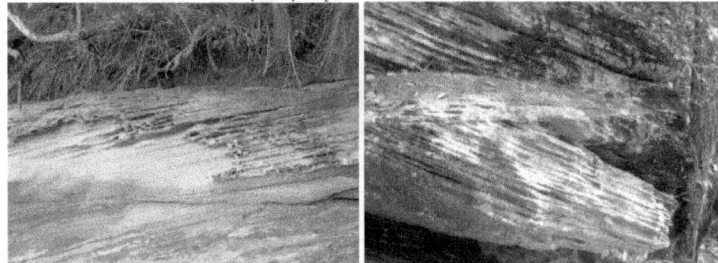

Left, honeycomb weathering in cross-bedded sandstone, Broken Bay, Australia. Right, Muogamarra sanctuary.

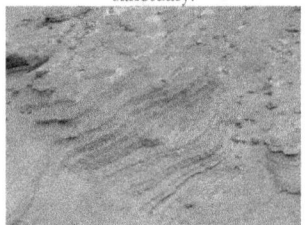

An eroded volcanic flank, Bartolomé, Galapagos Islands, beds that really *were* laid down on a slope.

Conglomerate.

Pebble beaches are a sure indication that the waves are rough, or the currents are strong—or both, delivering rushing water that carries off the sand and silt.

Cobble and pebble beaches between Gerringong and Kiama, NSW.

The same reasoning applies when you see conglomerate: wherever it is, it was always laid down by strong currents.

Permian conglomerate, Budawang Ranges, NSW.

This particular conglomerate, by the way, plays a major role in a story, later in this book. It lies at the very base of the sediments that make up the Sydney Basin at this point. It clearly formed when a huge flood pushed pebbles and boulders into an area that had been dry land, but as you can see in the photo below, there was other sediment in the mix, and we can see some mudstone across the lower middle. Mostly, it is pebbles, all the way through.

Permian conglomerate, Budawang Ranges, NSW. Note the mudstone on the left.

Thick beds of conglomerate like that are annoyingly rare, annoying because we humans like theories that involve catastrophes. Our mythologies are full of flood legends from Atlantis to Gilgamesh to Noah, our cinemas are full of asteroids, alien invasions, ships sinking and monsters arising from the deep. Most conglomerate beds came from minor floods that would never have made it into the 6 pm news. Here are two news-makers and two trickles:

Pebbly conglomerate in a road cutting, Manly NSW and in a thin bed, Blue Mountains, NSW (right).

Limestone.

Many limestone sediments contain the skeletons of microscopic sea life that settled at the bottom of an ancient sea. The shells and skeletons would yield a poor soil, indeed, but these deposits also contain clay particles which result from erosion on the continents, and maybe even small amounts of dust that landed in the oceans and then settled out.

Near Škocjan caves, Slovenia, a stream cuts down into the limestone.

Limestone is unusual, because it produces its own landform, karst topography. That means a region with irregular limestone strata riddled by hidden streams, running into and through sinks, caves and other subterranean passages.

To be strictly scientific, slightly acidic water reacts with the limestone, stripping it away. In Slovenia, near the Škocjan caves, the acidic water has torn out steep gorges like the one shown above, and even today, the streams are still cutting the channels deeper.

One of the more spectacular examples of limestone country is in County Clare in Ireland, in an area called 'The Burren'. The name is derived from the Irish name, *Boireann*, meaning "a rocky place", and according to tradition (meaning everybody quotes it, but nobody knows the source), one of Cromwell's army said of the place in 1651:

> ...of this barony it is said that it is a country where there is not water enough to drown a man, wood enough to hang one, nor earth enough to bury them. This last is so scarce that the inhabitants steal it from one another and yet their cattle are very fat. The grass grows in tufts of earth of two or three foot square which lies between the limestone rocks and is very sweet and nourishing.

The area strikes the visitor with its lack of soil cover and the extent of exposed limestone pavement. All the same, the gaps between the stones offer a rich mix of floral species, but it remains a rocky and windswept place better suited to tombs than to life of any sort.

Poulnabrone dolmen in Ireland, a tomb dating back to between 4200 and 2900 BCE.

The Burren plateau features subterranean drainage systems that are typical of limestone terrains. Streams disappear underground at swallow holes, and the waters flow through cave systems before emerging from springs, far away.

Limestone pavement, The Burren, Ireland.

Limestone pavements like the Burren are a legacy of an ice age that ended 15,000 years ago in Ireland. The glaciers ripped out the surface soil, stones and weathered rock, leaving a massive, uneroded rock surface. These forms are common enough where there is limestone in the high Alpine areas of Europe.

On the surface, the limestone landscape remains as an apparent stony chaos to the casual observer, but the bare rocks of the Burren are surrounded by rich patches where soil has blown in, and water holds, for now. These crevices are on joints in the limestone which catch soil and dead plant material, creating an acidic bog that eats down through the rock until it cuts a way to an underground cave.

Škocjan caves, Slovenia, near the exit, with stalactites. Note the tiny person at the base for a scale.

Sooner or later, the water conditions change, and the HCO_3^- ions come out of solution, combining with calcium ions to form insoluble calcium carbonate again. In some cases, this happens at the lower end of a downward drip, and a stalactite forms as the water evaporates near the lower end of the drip.

Under tropical conditions, stalactites can even form in the open, as seen in the next picture, which shows limestone formed from an uplifted coral reef in Vanuatu. Look at the marked area at the top of the exposed rock face. The rainfall here is about 1600 millimetres a year

External stalactites near Fels Cave, Chief Roi Mata's Domain, Lelepa, Vanuatu.

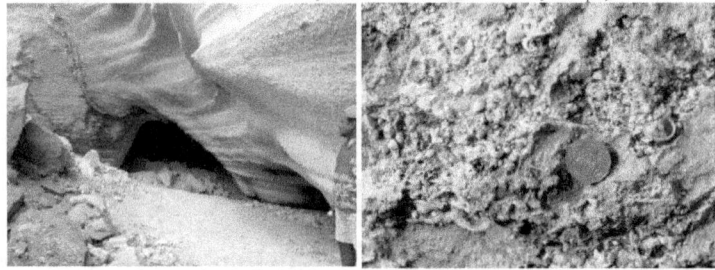
Fels cave, Vanuatu (left) and a shell-rich limestone formed from a shell bank, Mandurah, W.A.

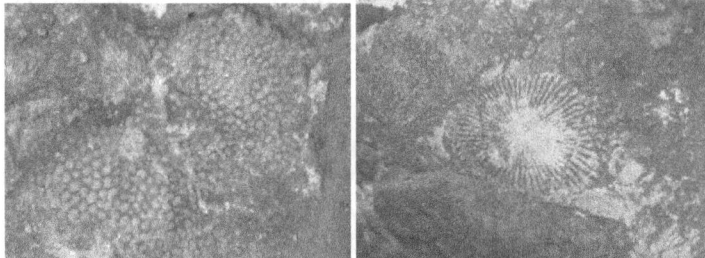
Devonian fossils, limestone floor, Wellington Caves, NSW.

Laying down beds.

> ...and a highly geological home-made cake.
> —Charles Dickens, *The Life and Adventures of Martin Chuzzlewit*, 1843, chapter 5.

Most of the sediment beds we see will never become sedimentary rock, because sooner or later, they will be disturbed. Still, at some time, somewhere, something happens similar to the way we can see temporary layers forming, but other layers settle in an area that remains undisturbed for a long time, probably because it is deep under water.

Careful observation of the rocks can often tell us about the conditions in which the rock was laid down. At the simplest level, fine sediments like mud and clay only settle in calm conditions, while you need a faster current to push sand along, and so on.

The photograph below was captured from a coach flashing by, but my interpretation is that we have mixed layers of shale and jointed sandstone, all horizontal, which means they must have been laid down under water.

Mixed sedimentary rocks, Zion National Park, Utah.

Mudstone, siltstone, claystone and shale.

Some geologists say the difference between mudstone and shale is that mudstones break into blocky pieces, while shales break into thin chips or layers with roughly parallel tops and bottoms, the splits running along bedding planes. That said, these are all sedimentary rocks made up of very fine sediment, mud, clay and silt, typically less than $1/16$ mm in diameter. They are mostly made up of clay minerals, and quartz and feldspars.

When they weather, shales tend to produce a gentler slope than other rocks, and among mixed sedimentary rocks, geologists would say the headland shown below is made up of shale beds between two layers of sandstone.

Narrabeen Series rocks near Box Head, Australia.

There is one very special shale, the Burgess Shale, which was laid down in the Middle Cambrian in British Columbia, Canada. These shale beds are remarkably fine, so the animals preserved in the stone are extremely detailed.

This shale was celebrated by Stephen Jay Gould in his book *A Wonderful Life*, in which Gould points out the huge range of phyla present in the shale, and suggests that, had things been slightly different, evolution could have favoured another group of animals in the shale, rather than supporting the descendants of a rather insignificant little beast called *Pikaia gracilens*, the earliest known chordate, which is either the ancestor of all of the vertebrate groups that we see alive today, or a close cousin of our ancestor. The rocks were laid down during the Cambrian explosion (chapter 10), when lots of new animal types were emerging.

This apparent record of sudden increases in variety may only be due to changes in ocean chemistry, or the numbers of predators in the seas, or some animals developing hard exoskeletons, remains that were easy to fossilise. It may also reflect a genuine increase in diversity, possibly brought about by the evolution of vision, sexual reproduction, or some other variation in animal life styles.

The existence of many varieties of animal in earlier deposits such as the Pre-Cambrian Ediacara formation makes many scientists suspect that the 'explosion' is just an apparent effect with no basis in historical fact, but it seems there was a lot going on when the Burgess shale was laid down.

Curiously, the Ediacaran faunas seem to have disappeared by the time the Cambrian rocks begin to give us evidence of the 'explosion'. Maybe the Ediacaran animals were still around, but their bodies were scavenged by the new animals, maybe they had all become extinct because of a catastrophe, maybe they were just out-competed by the new animals.

The most extreme viewpoint, taken by Adolf Seilacher, is that the Ediacaran faunas, even though found in many parts of the world, are just artefacts of some sort, that they are not true fossils at all. He points to the rocks involved, coarse-grained sediments formed in shallow turbulent waters, asking if this is the sort of place where we would expect fossils to form?

Similar deposits are known from other parts of the world, with about 30 recognized sites, mainly in Greenland, south China and north America, but none compares with the Burgess shale.

Faults.

> FAULT, in the language of miners, is the sudden interruption of the continuity of strata in the same plane, accompanied by a crack or fissure varying in width from a mere line to several feet, which is generally filled with broken stone, clay, &c., and such a displacement that the separated portions of the once continuous strata occupy different levels.
> —Charles Lyell, *Principles of Geology*, volume 3, p. 68

Normal fault in varved shale, Sandy Hollow, Hunter Valley, Australia.

Everybody has their faults, and I keep mine in a small block of varved shale that sits on my desk. I removed the things that sit on it for a photograph (left), but to get a clearer shot, I then put the other side down on a scanner (right image).

In the left-hand picture above, you can see the main fault, slanting to the left down the face of the rock, and the strata on the left have slumped down, which makes this a normal fault, said to be due to tension. If the left side had moved up under compression, it would be a reverse fault. Careful examination suggests that this fault arose from localised slumping: there are several other faults in the block, and there is folding at the bottom.

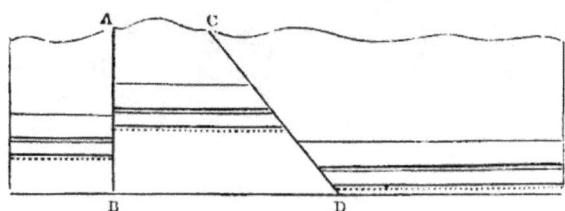

Faults. A B perpendicular, C D oblique to the horizon.

From Lyell's *Students Geology*, published before normal and reverse faults were identified.

In a normal fault, material to the left of the / symbol (or to the right of a fault sloping like this: \) moves down. In a reverse fault, the rock to the left of / or to the right of \ moves up instead.

When there are two nearby faults like / \ under compression, the rock in the middle slips down, in what is called a graben, while under tension, the block moves up to form a horst. When there are two nearby faults like \ / under tension, a graben forms, while when they are under compression, a horst forms.

> Handy mnemonic: a horst looks like a vaulting horse, the graben is a horst of a different colour

Larger-scale faults occur as a result of major force, but pictures of these are generally hard to come by and unconvincing.

Reverse fault, Camel Rock, NSW, Australia and normal fault, Waitemata sandstone, New Zealand.

Reverse fault, eroded volcano, Bartolomé, Galápagos Islands (close-up view on right).

Jointing.

Joints are, as explained earlier, fracture planes in rocks, and they are blamed on stresses that occurred, usually assumed to be the result of decompression when the erosion removes a rock's overburden, reducing the pressure below.

Against that, joints seem to go an awfully long way down, but for want of a better explanation, we probably have to stick with that for now. These planes of weakness are important for a number of reasons, one being that water can pass up, down and along joints, sometimes depositing solid material in the joint, making the joint more resistant to weathering and erosion.

Jointed Waitemata sandstone, Auckland New Zealand (left), jointed sandstone, Wangi Wangi, north of Sydney, Australia (right), each with iron deposits.

At other times, igneous dykes can intrude along joints, but one of the major geological effects is that weathering intrudes into solid rock along the joint lines, and these deposits can sometimes be combined with Liesegang rings, which were described in chapter 3.

As you can see in the next few pictures, and also in the Wangi Wangi rock platform above, iron concretions are important in shaping the surface of sandstone.

In the left-hand picture below, we see Liesegang effects sculpted into 3D, but the long, skinny concretion on the right is more amusing. It was, at one point, claimed by young-Earth creationists to be "incontrovertible evidence" of Noah's flood, because they mistook the concretion for a tree trunk.

Two effects of iron salts, carried into sandstone by water, Box Head, north of Sydney, Australia. The climber has his back to the 'tree'.

The Elvina gang, of which I am a part, examined the "tree" and showed it to be concretions passing through beds which have left their marks (like the cross-bedding near the top). The creationists quickly dropped their claim and ran off to find something else to misrepresent.

The bottom line is that joints are breaks in rocks or minerals, generating two free surfaces where none existed before. They offer planes of weakness that housing, road, bridge, tunnel and dam builders need to take into account.

Joints, Hawkesbury sandstone, Sydney Harbour.

Joints often come in regular sets, all trending in the same direction, and these are usually seen as a set, and two sets in the same rock form a system. In the photo above, the main set is trending north-south, out to sea, but the careful observers will see a second family of joints, running at right-angles to the first set.

Inner North Head, Sydney, Australia, showing joints in the sandstone. Blocks undercut by weathering plummet down, leaving a vertical rock face.

Road cuttings can reveal a lot about joints, showing the horizontal joints that are usually hidden. The horizontal planes show up clearly in the next photograph, and so do the two intersecting sets of vertical joints. This cutting was probably made in the 1920s.

Hawkesbury sandstone with three sets of joints (one horizontal), Fairlight, Sydney, Australia.

The photograph above reminds us that textbook diagrams are often deficient. Just above the base bed, there is an area of rock that masons would not use. This thin lens of shale probably represents a break in the sequence of deposition.

Joints are most easily seen where rock is exposed to the sea.

Folding rocks.

Like a teenager's bedroom (or the study, wherein I write), folded rocks are severely offensive to the ordered mind. This is why no ordered mind should ever be allowed to step up to the position of World Dominator (a term taken to encompass World Dominatrix, this being an equal opportunity work), because an ordered mind, given the power, would wipe out all the wiggly bits, and that would be a shame.

Folds visible on Cascade Mountain, near Banff, British Columbia.

Under very high pressure, many solids such as rock and ice become viscous fluids that flow. Folded strata appear, at first glance, unnatural and definitely not right in any way, but when you start to look around, they turn out to be common.

This boulder from the kink zone at Mystery Bay, NSW appeared at the start of this book.

In the 19th century, a standard "strong man" act always involved bending an iron bar. It requires a great deal more force to bend rocks than bars of iron, and this power usually comes as compressional or shearing forces when plates of rock come in contact with each other. The rocks at Mystery Bay are Ordovician, almost 500 million years old, but the mystery of the bay's name relates to an apparent murder of four men in 1880, and the geology is quite straightforward.

The kink zone is 400 metres wide, and in technical language, shows up in coastal outcrops of Ordovician turbidites (chert, black mudstone and slate). At some stage in the past, this area was the target of huge forces from a massive collision between plates, during which time the rocks squished like toothpaste.

Charles Lyell explained folding by relating it to a number of layers of cloth being squeezed between books, but the real forces are far greater than just being squeezed by a couple of slim volumes.

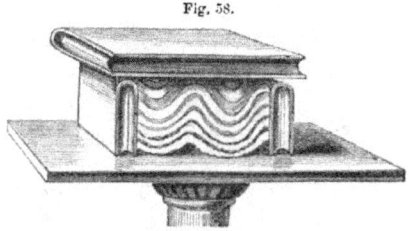

Lyell's model to explain "curved strata" in *The Student's Elements of Geology*, 1871, 54.

Not too far south of Mystery Bay, and still among the Ordovician rocks, is the Aragunnu campground in the Mimosa Rocks National Park. This is accessed from a dirt road mainly used by people who assume nobody else is there, so drive slowly.

The formation in the next picture doesn't seem to be mentioned in the tourist literature, even though they mention other things about the geology, so you will need to go down onto the beach and look around for yourself, if you want to find this amazing collection of folds and faults, but as to the nature of this rock, the deponent knoweth not. Here is what you are seeking:

Faults and folds in Ordovician rock, Aragunnu Beach, NSW.

In Switzerland, Mt Pilatus looks down on Luzern, and tourists like it for the views. Those who know their geology know that this is one of the rare places in the world where older rocks are on top of younger rocks. This is because tectonic forces became involved, as Africa pushed north into Europe. These forces left their biggest mark in the Alps, where the rocks are piled higher and deeper.

Folded limestone, top of Mt Pilatus, Switzerland.

Sometimes, something just has to give, and that usually means some rocks being folded. In this case, there was a thrust fault, and that has produced overfolding and a reversal of the usual age order. Steno's rules have not been broken, just run over. We do need to adjust the rules a bit, sometimes.

A few geological terms are best learned by knowing some entomology, rather than etymology. For example, stalagmites and stalactites may be recalled by "ants in the pants: the mites go up, and the tights go down". In a similar way, an anticline in cross section, looks a bit like a cartoon ant hill, while a syncline in cross section resembles a sink.

An anticline, partly exposed in a sheep paddock, NSW Australia.

Between Yass and Wee Jasper in New South Wales, some paddocks of a sheep station lie close to the road. In those paddocks, there is an anticline. It shows up faintly, at points where the rock at a given level is more resistant to weathering, so it pokes through the grass, tracing the shape of the geology below.

In western Maryland in the USA, there is a syncline mountain called Sideling Hill. At first glance, a "syncline mountain" seems odd: you would expect a valley, but a mountain is what you get. My thanks to an old friend, Laura Hicks of Kentucky, for these shots, taken on the fly:

Sideling Hill, Maryland. [Laura Hicks].

And here is a chance shot, taken from a coach window, while whizzing through the countryside in Ecuador. This same glacial till features at the end of chapter 3.

Folding in glacial till near Quito, Ecuador.

Sometimes, you may see a fold which is nothing of the sort, and any gully or ridge in flat and tilted rocks can give a false impression, as Charles Lyell knew:

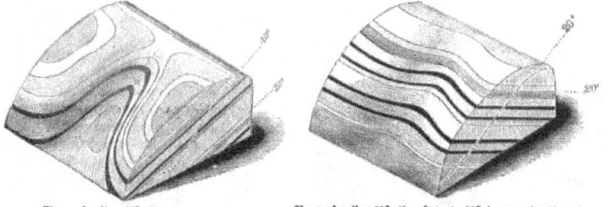

Slope of valley 40°, dip of strata 20°.   Slope of valley 20°, dip of strata 20°, in opposite directions.
Lyell's model to explain false folds in *The Student's Elements of Geology*, 1871, 61.

Several years ago, I spotted the apparent double fold seen in the left-hand picture below, but I went along Lake Como by boat until I got the perfect alignment, showing that the light-coloured bed was entirely flat: the folding was an illusion.

Looking from Bellagio, across Lake Como, Italy, four different angles.

Tilting rocks.

Some beds, as we see them, are simply raised up at one end. There may have been folding or even a total fracture at some other point, but we are now left with just a slope to look at. Don't worry about causes, just enjoy them!

Mt Edith Cavell, B.C., Canada (left) and Barragoot Beach, NSW, Australia.

137

Karekare Beach, New Zealand (left) and the south coast of Croatia.

Unidentified mountain near Plitvice, Croatia.

Bouncing rocks.

Various evidence tells us about the earth's inner structure, although the best evidence comes from the way earthquake waves are transmitted through the planet. The Earth has a crust on the surface, a mantle below the crust, and a core at the very centre. The crust is less dense than the mantle, and floats on it.

As we saw in chapter 2, the Earth's crust and the mantle are separated by a division called the Mohorovičić discontinuity, the point where the density of the planet changes. At continental fringes, the continental shelf joins the continental slope that ends in the continental rise near the abyssal plain, and the Mohorovičić discontinuity is closer to the planet's surface. This is why the Mohole (look it up!) was drilled in deep ocean waters.

Isostasy is the name given to the theory that the Earth's crust can be considered as less dense blocks floating on the denser semi-molten mantle. High mountains in this model must be regions where the crust is thickest, with deep roots extending down into the mantle, and this is borne out by measurements.

Follow the arrow and notice how the level rises when the weight of froth is taken off this pint of Guinness.

Following on from this, continents rise by isostatic uplift or elastic rebound when material is removed by erosion or an overburden of ice is melted away, as at the end of an ice age. This movement due to isostasy is quite slow, and much of Scandinavia is still rising up out of the earth as it rebounds from the pressure of huge glaciers during the last ice age.

Isostatic effects in the earth include ice loading in ice ages and the later rebound, as the previously loaded rocks are able to rise back up from the mantle. Mountains are a part of the less dense crust that floats on the more dense mantle. Mountain 'roots' sink deeper into the mantle.

Glacial rebound operates when glaciers start retreating, losing ice from the top. Scandinavia has fewer worries than most parts of the world from rising sea levels, because the land there is still rising out of the sea. Imagine an elephant sitting on a surfboard: the surfboard sinks, but bounces back when the elephant gets off. This is what is happening with Scandinavia, but more slowly. Maybe we should imagine a surfboard in honey?

Just south of Stockholm an island called Djurgården was separated from the mainland by a narrow channel along the north side of the island, but as the land rose, the channel got shallower and shallower. Then between 1825 and 1834, the channel was dug out and deepened, to counter the effects of glacial rebound.

The land continues to rise by about one centimetre each year, so sooner or later, the canal will need to be excavated again. Still, if glacial rebound causes problems for engineers, it is wonderful for archaeologists and historians. In

2016, I examined some amazing petroglyphs—rock engravings—at Alta in Norway. These are on bare rock, close to the sea.

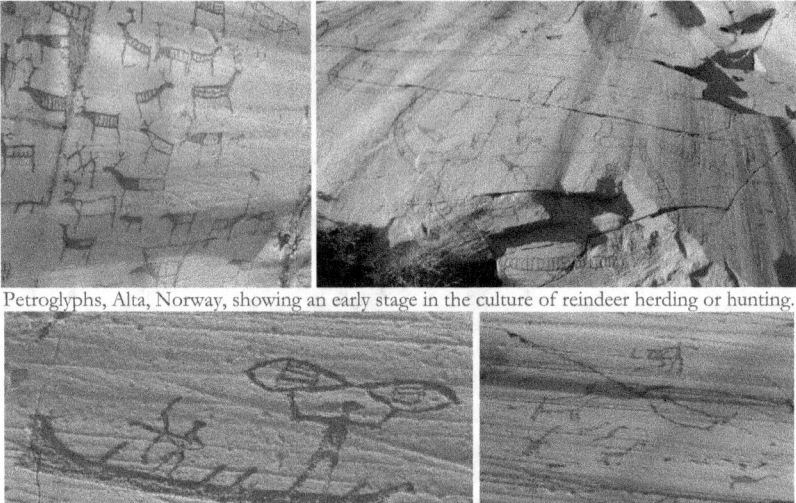

Petroglyphs, Alta, Norway, showing an early stage in the culture of reindeer herding or hunting.

Other scenes show fishing, and possibly hunting.

These engravings were probably all done close to the water's edge, meaning the highest are the oldest. When something like a railway line appears (second photo) this probably shows a time when the local people started making fences around their reindeer.

In November 2016, it was estimated that 90 per cent of the rise in the Swiss Alps since the last ice age was caused by rebound, as the Alpine glaciers melted. If a new ice age comes, the glaciers will slowly advance once more, and the land will be pushed down again. Right now, a new ice age seems unlikely, but we have a less than complete understanding of climate, so who can tell?

And now, a small puzzle:

Is this a leg-pull?

There is a totally counter-intuitive power source on the island of Kefalonia in Greece. Near today's principal town, Argostoli, the rock is limestone, and at certain points, there are swallow holes (*katavothres*), into which seawater flows. That is hard enough to accept, but the use which was made of the effect sounds like the sort of hoax I sometimes pull. I tend to believe this one, though…

Geologists have now explained that the seawater flows in, travels under the island, and driven along by freshwater currents, the water is carried out, fifteen days later, into the sea on the other side of the island at Melissani Cave. In

1865, the Milliaressi family decided to make use of the currents to drive a flour mill.

The seamill near Argostoli, Kefalonia.

They called it a *thalassomoulos*, or seamill, but to this day, visitors suspect that their legs are being pulled. This suspicion is heightened when they see that only a few traces are left, along with a non-working reconstruction, further along the shore, designed to attract tourists to a rather seedy cafe.

The modern reproduction of the seamill near Argostoli, Kefalonia.

This particular energy source is (probably) real, and it was once quite famous. At the same time, the readers is cautioned that, just because one weird scheme actually works, it doesn't mean the other schemes will!

6: Changing rocks.

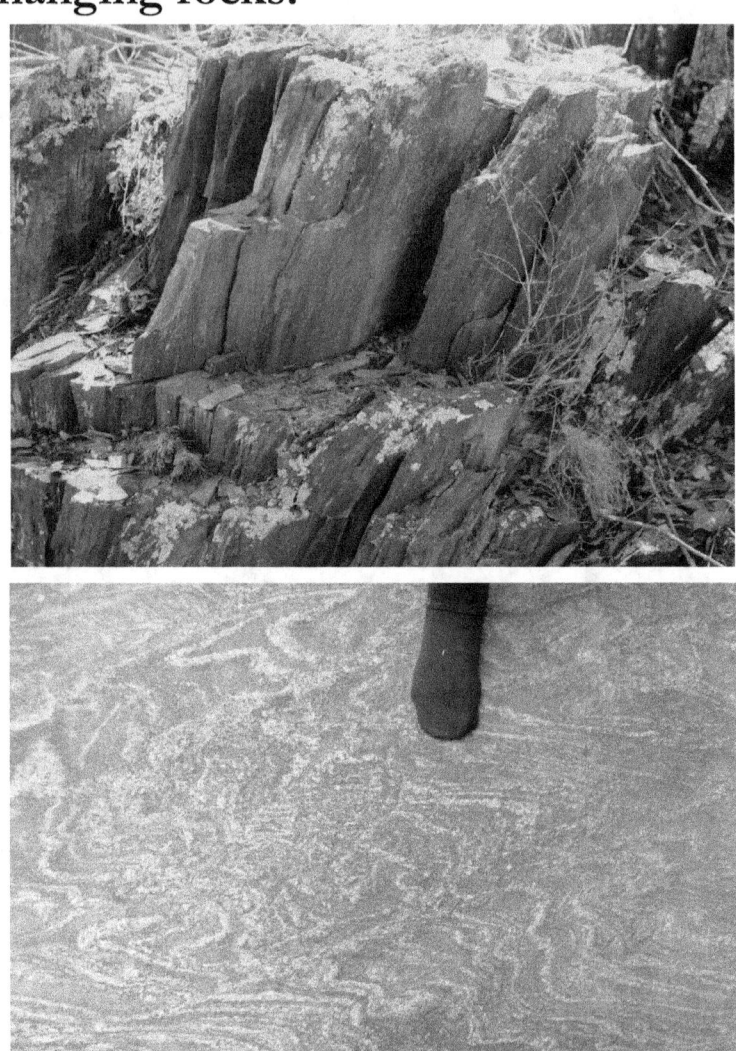

Weathered slate, Apsley Falls, NSW; Gneiss floor of a Sri Lankan cave, with a socked foot as a scale.

Metamorphic rocks have been changed by a combination of extreme heat and extreme pressure. These rocks were once something different, but they changed their form, deep below the surface. This is going to be the shortest chapter in this book, because, while metamorphism is interesting enough, most of the rocks are fairly hard to understand.

Examples include gneiss, slate, marble, quartzite and schist. There are two main kinds of metamorphism, regional and contact metamorphism, as well as burial metamorphism. While pressure plays a role, heat is extremely important as a way of "softening up" the rock.

In contact metamorphism, there is no real pressure involved. This is the effect when a dyke, sill or flow causes local heating of the pre-existing rocks with which the hot rock has made contact. Unlike regional metamorphism, this can happen at shallow depths, it operates over a small distance.

This is what happened in the North Bondi case, pictured here, where the evidence is in the slanted columns. When a volcanic neck erupted through the sandstone on what is now a cliff north of Bondi Beach, the sandstone was melted, resulting in columnar jointed quartzite.

 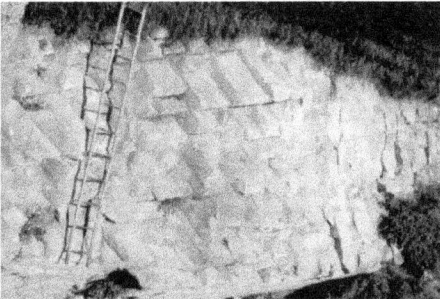

Contact metamorphism, North Bondi, NSW, Australia. You have to cross a golf course to get there.

The crystals in a metamorphic rock can be chemically the same as in the original rock, but they will usually be larger and interlocking. As an example, sandstone made of quartz grains held together by a silica cement will become quartzite, with large interlocking silica crystals.

If the original rock contained more than one mineral, the squishing effects of pressure push similar particles into aligned bands of those same minerals, but it is possible for new minerals like garnet to form as well. Keeping it simple, there are two types of foliated metamorphic rock that are easy to spot.

Schist: this sort of rock has a flaky appearance due to drawn-out bits of mica.

Slate: this kind of rock has no visible banding at all but it shows rock cleavage. Slate is often split into thin sheets and used as roofing material, but in the 19th century, school children learned to write on slate.

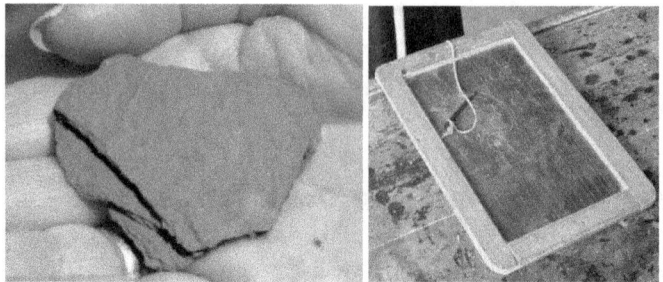

Slate, eastern Switzerland; school slate in a Norwegian museum, Maihaugen.

There are two main non-foliated metamorphic rocks with (officially) no banding evident:

Marble: made of calcium carbonate (limestone), fused together.

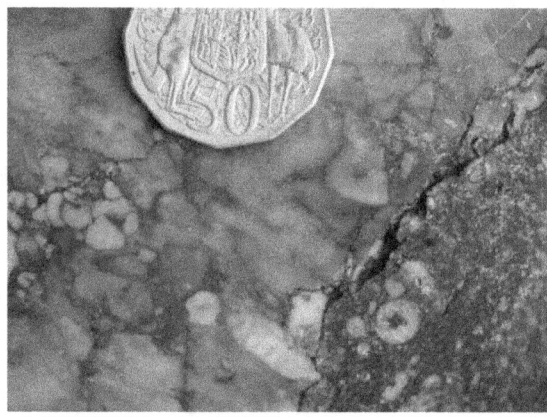

Marble table showing fossils. The coin is 32 mm across. As explained in chapter 10, it is my standard scale.

Quartzite: made of fused silica.

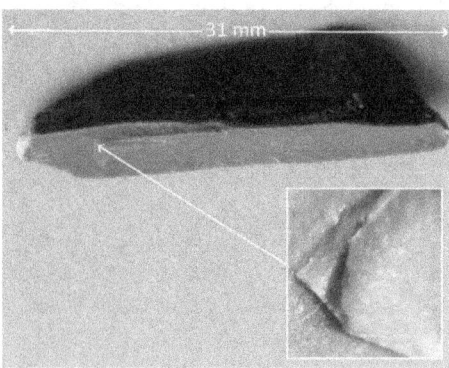

Quartzite blade made from a pebble by the author, Chefchaouen, Morocco.

Where does the heat come from?

The planet gets hotter as we dig deeper. The Beatrix mine in the Witwatersrand Basin in South Africa reaches 2200 metres below ground, and the temperature increases by about 29°C for each kilometre. At the bottom of the mine, the temperature of the rock is close to the boiling point of water on the surface. The air pressure is almost 30% more than it is on the surface.

At the bottom of the mine, the heat is reduced to safe levels by ventilating the mine with cool air, but if you went down another 10 km, the rock temperature would be approaching 400°C, the air pressure would be huge, and the downward pressure of the rock would be much higher.

When you dive 10 metres below the ocean's surface, the pressure of the water is as great as all of the atmosphere's pressure, and we say the total is two atmospheres, 30 psi or 200 kilopascals, about the same pressure as the air in the tyres of your family car. That's one atmosphere from air and one from water.

Just 3 metres of rock exerts about the same pressure as 10 metres of water, and at the bottom of the mine, the downward pressure from the rock above is more than 70 megapascals. If you were to drill down 10 km below the bottom of the Beatrix mine, the pressure would be more than 300 megapascals (MPa), or 3000 atmospheres.

> Remember that elephant on the surfboard? Imagine an African bush elephant (the biggest type) wearing stiletto heels, balancing on just one of those heels (look, I said *imagine*, right?). The pressure under that heel would be around 300 MPa.

Go down another 12 km into the planet, and the heat and pressure can make rocks do strange things. They squish, the crystals in the rock get rearranged and the rocks become metamorphic rocks.

Undergoing metamorphosis is probably the ultimate fate of all minerals, though they may go through several cycles as sedimentary rocks first. Sir John Herschel, an astronomer, wrote a letter about this to Roderick Murchison, in 1836.

> In the formation of these therefore there is nothing casual [in the formation of metamorphic rocks]; all strata once buried deep enough, and due time allowed, must assume that state. None can escape; all records of former worlds must ultimately perish.
> —Quoted in *Arcana of Science and Art* (1838), 258.

In 1835, Charles Lyell pointed to the heat at lower depths as a key element. In 1830, he said, a Mr Fox had established that warm air came out of deep mines in Cornwall, heated by 9 to 17°F (5 to 10°C).

> If we adopt M. Cordier's estimate of 1°F for every 45 feet of depth as the mean result, and assume, with the advocates of central fluidity, that the increasing temperature is continued downwards, we should reach the ordinary boiling point of water at about two

miles below the surface, and at the depth of about twenty-four miles should arrive at the melting point of iron, a heat sufficient to fuse almost every known substance. The temperature of melted iron was estimated at 21,000°F, by Wedgwood; but his pyrometer gives, as is now demonstrated, very erroneous results. Professor Daniell ascertained that the point of fusion [of iron] is 2,786°F.
—Charles Lyell, *Principles of Geology*, volume 2, 357–8.

T
How do you measure the temperature at heats where any normal thermometer melts? That was the problem Josiah Wedgwood (the potter named above) faced. He was a mass manufacturer of pottery, and so he invented his pyrometer. Keep in mind that what happens to the clay in a pottery kiln is like metamorphosis in rocks, so it is worth our while diverting to look at this.

We know the name Wedgwood now as a brand of high-class pottery, and we think of his creamware and his Jasper ware, stuff intended for the nobs and nabobs. These high-quality items were useful selling points, but he actually made his money from the cheap mass-produced white earthenware pottery that even the working classes could afford.

He ensured quality by using his pyrometer: if you use mass-production, you need quality control. The craftsman is replaced by an unskilled worker who shoves batches of pots into a large kiln, somebody who needs to be guided.

Clay expands when it is heated: the hotter it gets, the more it expands. So if you take a standard test piece and heat it, the size it expands to will depend on the temperature of the kiln. If you measure the test piece, you can see how hot the kiln was. That sounds easy enough, but the differences in size are very small.

Wedgwood made a sliding scale, two pieces of metal which taper towards each other, ever so slightly. Put the standard test piece in, slide it along, and see where it stops sliding, and read off the kiln's temperature. It was simple, and it worked, and everybody was impressed. They even made him a Fellow of the Royal Society for his ingenuity.

Even more impressive, the potter had a wooden leg, and when he found a piece of sub-standard pottery in his works, he would smash it against his leg.

Just as a totally side-issue, Wedgwood's black pottery was perfect as a streak plate for the acid test used to investigate gold, or more particularly, faked gold, which is an alloy of another less valuable metal with some gold.

Gold is one of the few metals which resists acid, so jewellers 'streak' the metal across a piece of black pottery, and then the streak can then be tested with acid. The more the streak disappears, the more cheap metal there must have been in the original streak.

Now back to metamorphic rocks, which have been changed by a combination of heat and pressure. In the case of contact metamorphism, heat is the main driver, and this often tells a story. There are two kinds of horizontal basalt layers that may be found, later, sandwiched between layers of sedimentary rock.

One is a flow, where liquid basalt reaches the surface and flows sideways over the top of the surrounding countryside, before sediments cover it and later become rock. The second form is an intrusion called a *sill*, which is a sheet-like body of intrusive igneous rock which fits between the bedding or structural planes of the host rocks. Here, liquid basalt rises upwards along a joint, but then pushes out sideways, between two layers of sedimentary rock.

If later sedimentary rock settles and forms over a flow, it can be hard to tell that it is not actually a sill, but the answer comes from checking for contact metamorphism in the overlying rocks, because there will be none in sediments which later fall on top of an old, cooled flow.

Sills and flows are important in relative dating, where all that matters is the relative age of a rock, rather than its absolute age. This involves correlating beds over large distances, or making assumptions based on the intrusions which have caused contact metamorphism.

This becomes important when we are looking at short gaps of time, of the order of tens of thousands of years, as in the study of hominin evolution. When fossils in two places lie under (or just over) the same flow, they are of similar age, but sills can be unreliable indicators.

Fossils in metamorphic rock.

A rock taken from Akilia, Greenland, has uncertain geological origins. Talking technicalities, it may be a banded iron formation in a metamorphosed sedimentary rock, but it could also be a metasomatized ultramafic igneous rock.

Don't worry about the names, but the rock's nature matters, because it may provide evidence of life on our planet some 3860 million years ago. There are no actual fossils as such in the rock, but the levels of carbon-13 (^{13}C) in the rocks have been interpreted as evidence that something living made a contribution to the material that made up the rocks, some 3860 million years ago. Science can be like that, a matter of reasoning and interpretation.

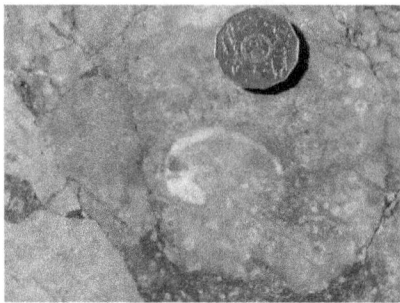 

The same marble table seen before, showing fossil traces.

Some metamorphosed sedimentary rocks, especially marble and slate, may contain fossils, but any life from 3860 million years ago would almost certainly have been too small to leave any visible traces, leaving the scientists to dig around in the chemical composition of the rocks.

If the Akilia rock can be shown to be of igneous origin, rather than deriving from a sedimentary rock, that probably knocks out any chance for these rocks being used as evidence of early life. It does not mean that there was not life then, just that this is not evidence for life having been around then.

^{13}C is one of the stable isotopes of carbon, so the process here is not one of 'carbon-dating'—rather, it is a matter of biochemical reactions that cause variations of ratios from usual levels, when certain products are examined.

The structure of key interest shows significant drops in the ^{13}C ratios, but that only stands up as evidence if the rock is sedimentary. Some geologists interpret the rock as a magnesium-rich volcanic rock that was repeatedly injected by quartz-rich fluids to produce the appearance of banding. The jury is still out.

If I were a better geologist, I could explain these veins. Seen near Cuttagee Lagoon, south coast NSW.

At other times, the picture is clear enough to be seen at a glance:

Partly metamorphosed shale/slate, near Wadahl, Norway.

In the example above, the original beds of varved shale are visible in the 'slaty' blocks I found being used for a retaining wall in Norway. A day later, I found another wall, where cross-bedding was visible in a bed that was about 100 mm from top to bottom:

Quartzite showing traces of cross-bedding, near Bergen, Norway.

7: Things going wrong.

Multiple faults in a roadside cutting near Quito (Ecuador).

It is the hardest thing in the world to frighten a mongoose, because he is eaten up from nose to tail with curiosity. The motto of all the mongoose family is, "Run and find out"; and Rikki-Tikki was a true mongoose.
—Rudyard Kipling, *Rikki-Tikki-Tavi*, 1893.

[Pliny] died a victim to his curiosity…He was in command of the Roman fleet at the time of the great eruption of Vesuvius, which destroyed Pompeii and Herculaneum. He landed in order to watch the upheaval, ventured too far, and was overwhelmed by the storm of falling ashes.
—William Dampier, *A History of Science* (1942) page 61.

Despite what my publishers often assume, I am not really an historian. I do, however, write about historical matters, because like any good scientist, I am kin to the mongoose. I want to know why things happened, and what made them happen as they did, and not differently. That is why I can argue that without the steam engine and the telegraph, Einstein may not have started the line of thought that led to special relativity.

It all began with a need to synchronise clocks, but nobody needed to do that until railways came in, and timetables were needed to make sure that up and down trains did not collide on the single tracks that were normal. (The single tracks between towns worked because train drivers would pull into sidings to let other trains pass. That is why they needed synchronised clocks and watches.)

So Einstein's thinking arose from the needs of steam train drivers, but until the telegraph was available, synchronisation was hard. Without telegraphy, Einstein may not have pursued the deep problems of time and space.

Probably he would have got there, because trains and telegraphs were helpful precursors, rather than what scholars back then designated as *sine qua non*, a Latin tag meaning without which, nothing. There were many essential precursors like the invention of calculus, and probably a few enabling technologies.

Most sciences depend on glass, for example. We use or used glass for windows for houses and lanterns, to make light bulbs and tubes, lenses for spectacles, telescopes, microscopes and surveyors' theodolites, laboratory glassware from test tube to reagent bottle to Petri dish, thermometers, barometers, cathode ray tubes, optic fibres and much more.

Most modern science could not exist without glass, but tourism also played a role. The idea of the Grand Tour began in the late 1600s, but as late as 1843, travelling from London to Rome took 21 weary days, though in 1860, steam ships and trains got you there in just two and a half days. In those earlier and slower days, when people went on the Grand Tour, they tried to tick as many boxes as possible, and after Pompeii was rediscovered in 1748, it went on the list of marvels to see. Seeing things like that helped educate people more widely.

Volcanoes.

View of the Isle of Cyclops in the Bay of Trezza.
(Drawn by Capt. Basil Hall, R.N.)
Captain Hall's drawing of Isole dei Ciclopi, on Sicily's east coast, with columnar jointing.

When Mount Vesuvius erupted in CE 79, there may have been no media coverage in our modern sense, but the event was widely studied and reported, and we know that Pliny the Elder died when he got too close to the volcano, which also destroyed Pompeii, Herculaneum, and a great deal more.

And now, a jump out of the timeline: in the late 1990s, when geochronologists dated the eruption of Mt Vesuvius, the one that buried Pompeii in CE 79, but if we know the date from historical accounts, why would anybody bother to try to determine the age of the volcanic debris that flowed forth in that eruption?

The answer is found in a single word: *calibration*. Accurate radiometric dating of young rocks is essential for scientists in many fields and where samples from the Holocene, the last ten thousand years are involved, the primary method has been radiocarbon dating.

Mt Vesuvius, before dawn.

The problem with relying on just one method is that there may be systematic errors, such as the slow seepage of newer carbon into old deposits, topping up the ^{14}C levels and giving us spuriously young ages for material which is more than 40,000 years old. Nobody can say if this is a problem or not, at least until an independent way of assessing ages can be used alongside the carbon method.

So the plan was to use the argon-40/argon-39 method to date material from a known historical event, giving us a second version of the dates that the record books have established. Argon dating tells us how long it is since lava solidified and trapped radioactive elements in its crystal lattices, setting the clock to zero. All you have to do is measure the amount of potassium-40, a radioactive isotope with a half-life of 1.25 billion years, against the concentration of its daughter product, argon-40.

The method has been around for many years, but recent major refinements have made it possible to detect extremely tiny amounts of argon, making it possible to calculate the age of younger rocks. Because the decay rate is so slow, quite a few years have to pass before the argon levels are high enough to allow accurate measurement, so until now, the youngest rocks dated this way were more than 5000 years old, and the accuracy was no better than 10%.

By heating samples of volcanic ash with a very precisely controlled laser in careful steps, the researchers were able to date the rock, then 1918 years old, to 1925 years, plus or minus 94 years, a remarkably accurate result. (Note that dates like this always come with a confidence level: there is a real but small chance that the rock is outside the range 1831 years to 2019 years.)

Back to our history, people who travelled on the Grand Tour or for other reasons went, visited, observed, and reported back on what they had seen. Sir William Hamilton was the British Ambassador to the Kingdom of Naples from 1764 to 1800, allowing him to follow his other callings, as antiquarian, archaeologist and vulcanologist.

While he was not a scientist, Sir William was well-placed to study the excavations which were happening at Pompeii, and the activity of Mount Vesuvius. He was also, unfortunately, living at a port where Horatio Nelson would call, so that his wife Emma would meet and fall in love with the dashing Admiral of many parts, not all of them in the same place.

In 1794, a 'shower of stones' was reported at Siena, at a time when most people did not believe that meteorites ever landed. Sir Joseph Banks, president of the Royal Society, thinking the cause might have been a volcanic explosion, wrote to Sir William, seeking more information, and Sir William replied:

> As we have proofs during the late eruption of a quantity of ashes of Vesuvius being carried to a greater distance than where the stones fell in the Sienese territory, might not the same ashes have been carried over the Sienese territory, and mixing with a stormy cloud, have been collected together just as hailstones are sometimes into lumps of ice…

Right or wrong, Hamilton and people like him, made the scientists of Britain more aware of volcanoes and vulcanism. Hamilton's other main contribution to this story came when he published a book of the pottery he had collected, believing it to originate in Etruria. This work inspired Josiah Wedgwood to imitate the style, when in fact the pottery was about as Etruscan as Sir William. Who cares? It was still beautiful, while glowing ignimbrite lacked glamour.

The sloping layers of this headland at Eden, NSW, are composed of ignimbrite.

Some pyroclastic flows form a rock called ignimbrite. These are like powder avalanches, but instead of being made of cool air and cold snow powder, the gases are fearfully hot, and the finely powdered volcanic ash is red hot.

Clouds of gas and ash hurtle down the slopes of the volcano, converting all life forms to instant char and cinders. The available data are sketchy, but the usual suggested speeds are above 80 km/h. Against that, the damage to trees done by the cloud at Mount St Helens seems to suggest something more like 480 km/h.

While there are older deposits of ignimbrite in the area of Vesuvius (like the 40,000-year-old Campanian ignimbrite), the deposits around Pompeii from the famous eruption are classed as tephra. This, however, is getting a bit too technical, and poets would probably call both rocks granite, in order to find a convincing rhyme for *planet*.

While lots of people have written about earthquakes, there are very few eye-witness accounts of volcanoes, mainly because, if you are close enough to see, you are close enough to die. Gaius Plinius Secundus, often called Pliny the Elder, was an enthusiastic student of nature, and also the commander of a Roman fleet.

When Mt Vesuvius erupted and buried Pompeii, Pliny had some Roman galleys launched and taken across the bay of Naples to rescue people, but while on shore, he collapsed and died.

According to his nephew, Pliny was killed by gases from the volcano (not ash), but it is more likely that he suffered a heart attack or a stroke, since none of his companions died at the same time.

Still, Pliny is honoured, perhaps falsely, by modern scientists as an early martyr of science, and people have been taking many risks to study volcanoes ever since. The volcanoes, for their part, have many ways of being lethal. Then there are the volcanic bombs, blobs of viscous lava, more than 64 mm in diameter, flung upwards during an eruption. As the lava flies up and then falls, it may be pulled into an aerodynamic shape, as the liquid rock cools and hardens.

At dusk on the island of Tana, Vanuatu, Mt Yasur's blobs of lava fly, cooling and becoming bombs.

The terminal velocity of a volcanic bomb is probably less than 300 km/h, but that is enough to do a lot of damage. Some of them end up 5 km from the volcano, suggesting that they took off at something more than 800 km/h. Still, we are lucky: Jupiter's moon Io has a volcano that throws stuff out at 3000 km/h!

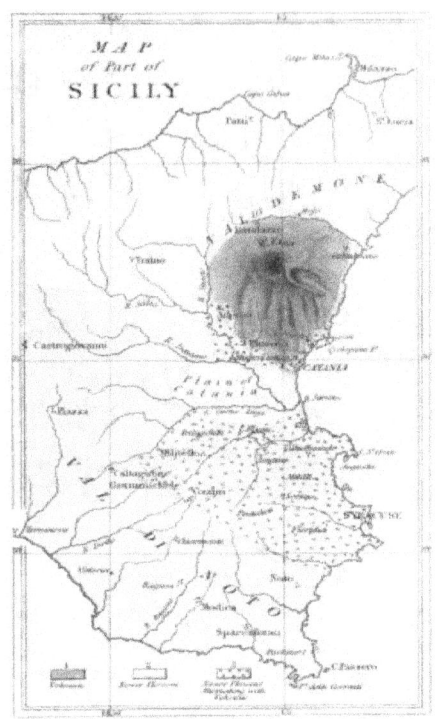

This map, from Lyell's 1834 third edition of *Principles of Geology*, shows Mt Etna on Sicily. The volcano covers 40 km from east to west, 50 km from north to south.

A giant Antarctic volcano.

There have been even bigger volcanoes in the past, and drilling in the late 1990s on the sea floor off Antarctica's Victoria Land coast near Cape Roberts, shows surprising evidence of enormous volcanic eruptions, some 25 million years ago.

The thickest distinct layer of volcanic debris is 1.2 metres thick, which suggests an eruption as dramatic as that of Krakatoa (also called, more correctly, Krakatau) in 1883. These layers contain volcanic pumice up to 1 cm in size, which suggests that the volcano was located within 50 to 100 kilometres of the drilling site and that it erupted in a style similar to Vesuvius.

The evidence points to a blast several times greater than the Mount St. Helens eruption, and possibly even comparable with the eruption of Vesuvius that destroyed Pompeii in 79 CE. These eruptions would have significantly altered global temperatures at the time. Mount Pinatubo, a much smaller event, cooled the world climate by 0.5°C for a year after its 1991 eruption.

Mt St Helens, as it appears now, after all of the near side blew out in the eruption.

The drilling in Antarctica was being done to get evidence about the climatic and geologic history of Antarctica during the last 100 million years. At a depth of 110 metres, the workers found unexpected evidence of volcanic activity in the form of layers of volcanic debris that had erupted explosively into the atmosphere, and then settled through the air, into the ocean and onto the seafloor.

The thickness and coarseness of the main debris layer indicates a large-volume eruption that generated an ash cloud that reached into the stratosphere. While volcanoes cause a great deal of local damage in the short term, they are very useful in the long term because they bring valuable new minerals to the surface. Seismic events, on the other hand, have fewer helpful features.

Earthquakes.

> ...and the mount of Olives shall cleave in the midst thereof toward the east and toward the west, and there shall be a very great valley; and half of the mountain shall remove toward the north, and half of it toward the south.
> —*Zechariah* 14:4

To those with even a little bit of geological knowledge, the prophet *must* have been writing about an earthquake. In fact, the very next verse refers to a previous earthquake, almost as if to reassure us.

This appears to be the earliest reference anywhere to movement along a fault line during an earthquake, although there had been earlier quakes in the same area. Quite a few people explain the collapse of the walls of Jericho by assuming an earthquake, and even the parting of the Red Sea for Moses and the Israelites is seen by some as the result of an earthquake.

An earthquake happens when local tension is released between two large blocks of rock pushing past each other, along a line of weakness. We will consider shortly just how and why the rocks move. For now, our interest lies in the lines of weakness, or, if you think in three dimensions, we are interested in the planes of weakness which are called faults.

What earthquakes do.

Earthquakes are very common, and instruments record some 800,000 in a normal year, but as we saw earlier, most of these are so small that we do not feel them. A severe earthquake, with a magnitude of greater than 8.0 on the Richter scale, can be expected every eight to ten years, but some smaller earthquakes can cause significant destruction each year.

An earthquake is a series of shock waves generated at a point (*focus*) within the Earth, caused by the movement of rocks on a fault plane releasing stored strain energy. The point on the surface of the Earth above the focus is the *epicentre*. Major earthquakes are associated with the edges of plates that make up the Earth's crust, and along mid-ocean ridges where new crust is forming.

The greatest concentration of earthquakes is in a belt around the Pacific Ocean (the 'ring of fire'), and along a zone from the Mediterranean Sea to the Himalayas and China. Major earthquakes, such as in San Francisco in 1906 and Japan in 1923, caused much damage to property and loss of life.

To most people, an earthquake means falling buildings, and victims trapped in the rubble, but the effects of an earthquake go far beyond damaged buildings. Roads are blocked, ports are closed, gas lines and petrol tanks fracture, causing terrible fires, and later, disease starts to spread among the survivors, huddled together in any large buildings that survived the earthquake.

There are other, less obvious dangers in an earthquake. In Kobe, Japan in 1995, influenza was on the increase, less than two weeks after the earthquake hit. The authorities acted quickly, and there were only a few hundred cases of influenza, but the risk was serious while it lasted.

Then there is the social disruption in an earthquake-prone city. While people are sleeping on stretchers in the schoolrooms, classes cannot take place, and family life is fragmented. The survivors in such cases begin to look for somebody to blame, and find ready scapegoats in the government, increasing their fury with every rumour.

In the Kobe case, they may have been correct to blame the government for some of their problems. Buildings collapsed which should not have fallen down, bureaucrats held up the entry of emergency supplies and even of rescue dogs flown from Europe. In the end, when the dogs were allowed into Japan they found just nine victims, all of them dead.

At first, there will only be a few roads open after an earthquake, and these will soon be clogged with traffic. The other roads will be filled with rubble, broken glass, fallen power lines, wrecked vehicles, and dead and dying victims.

Ambulances will be unable to reach most parts of a city, after an earthquake. Those lucky enough to reach a hospital will find little benefit, for modern hospitals rely on electricity. With the power lines down, and the emergency generators running out of fuel, there will be little the hospitals can do.

Ordinary citizens can get involved by moving in, where it is safe, and clearing roads so emergency vehicles can get in and out. After an earthquake, water, gas and electricity are off, and sewer lines are broken, so pit toilets are essential, and as we will see in chapter 9, that means you won't be able to drink well water. People will not even have clean water to bathe victims' wounds.

There are advantages for a few people: with phone lines and roads cut, the helicopter pilots and sellers of mobile phones will do well, but many other workers will discover their place of work is to be closed for months. Geology can intrude our lives in unexpectedly rough ways.

In Kobe, Koreans were the main workers in the shoe industry, but most of the shoe factories were destroyed in the earthquake, so their community was hit hard. Those people who were still employed found they were doing little business, and food trucks could not enter the city, so they could not buy food.

The Kobe earthquake was the third in the area in as many months that was over 7.0 on the Richter scale. It killed over 5,000 people, while an earlier July 1993 earthquake in the area only killed 200 people, even though it measured 7.8 on the same scale. Most of those 200 were killed by tsunamis. The Kobe earthquake produced no tsunamis, mainly because the focus was some 20 km below the Inland Sea, and very close to Kobe.

On the other hand, a small earthquake in Japan in 1896, one which hardly even rattled any buildings, killed 22,000 from tsunamis, and as we have already seen, a 1960 earthquake near Concepción in Chile caused a tsunami that killed 60 people in Hawaii, and 200 more in Japan. The whole of the Pacific Ocean is at risk from earthquakes elsewhere in the ocean, but earthquakes in the Aleutian Islands, near Alaska are considered to be very effective generators of tsunamis.

One of the main problems is liquefaction, where soil and underground water mix to form something like quicksand. Most of the Japanese building standards assumed that liquefaction is mostly on the surface, but in Kobe, it reached as deep as 10 metres on reclaimed land.

The damage in Mexico City in 1985 was also related to the large areas of reclaimed land in the city. Tokyo's waterfront is lined with chemical factories, oil refineries and densely populated neighbourhoods that sit on similar ground, so when Tokyo has its next large earthquake, those decisions will add to the extent of the disaster.

Over the centuries, Japanese architecture had adapted to the conditions, with light buildings that were easily destroyed in typhoons or earthquakes but could be repaired and rebuilt just as easily. Modern Kobe, like any other city of its age, contained buildings which fell just as easily, but which fell much harder, and were also much harder to replace.

Earthquakes: a personal note.

Our daughter, Associate Professor Cate Macinnis-Ng, is a plant scientist in New Zealand. She was in Christchurch on a fateful day in 2011. With two scientist parents, she knows how to observe, and here is how she described it for me.

> We heard it first as a low rumbling like the sound of a big truck approaching. "It's just an aftershock," my colleague reassured me as his office began to shake.
>
> I was in Christchurch for some meetings and was discussing a possible new collaboration with another plant scientist. Matt's office was on the third floor at the University of

Canterbury and he had experienced plenty of aftershocks since the first earthquake a few months ago. But this was different.

As noise got louder and the room shook from side to side more and more, Matt realised this was more than he expected. "Get on the floor!" he shouted above the roar as his books started falling out of the shelves and his filing cabinets started to open.

By this time, our desk chairs were rolling around and as the computer fell off the desk, I had the horrible feeling that the whole building would collapse. The sensation of jerking from side to side was no better huddling on the floor as more books and papers landed around us.

There was nowhere safe as the whole building was shaken about. When the noise and movement finally subsided, the rumbling was replaced with alarms sounding. Leaving Matt's trashed office, he noted sadly that he'd only just finished cleaning up after the last big shake.

Students and staff streamed down fire stairs and out to evacuation areas but I noticed all of the buildings were still standing and I felt a bit silly for thinking our building would go down.

As we walked across an outdoor parking area to leave the university, an aftershock hit and the cars started bobbing up and down as the earthquake caused ripples across the concreted area.

I sent a quick text to my husband and parents before the mobile phones cut off to let them know I was fine. With the university closed, we evacuated to the nearby home of another colleague while everyone else tried to get through the jammed roads to family.

With each aftershock, we could hear it coming, and we weren't sure if it was going to build into something big or dwindle before the shaking really started. The constant tension of aftershocks coming was immensely stressful. It wasn't until later when the power came back on and we were able to watch the television that I realised the devastation the quake had caused in the centre of town.

Our building was never going to collapse, but I was in the best part of town because the buildings at the university had been built well above the required standard. It was distressing to be within kilometres of so much destruction and heart ache.

I was grateful to be able to return to the safety of Auckland the next day. I wasn't the only one keen to get out of Christchurch, and the airport was packed with people trying to get on a plane. Passengers erupted into a round of applause as we took off from the runway.

It took several weeks for me to trust that the ground was solid again and the sound of trucks approaching will always have a different feeling for me now. I've been back to Christchurch several times since, and the city is coming back stronger than ever but the February 2011 earthquakes will forever be a big part of the history of New Zealand's third largest city.

In early 2020, I visited Christchurch to assess the social effects, nine years on, and learned there were many other aspects to consider. The water table (chapter 9) was close to the surface, and the vibrations caused a lot of liquefaction: anybody visiting the city is advised to visit Quake City, a museum which explains the science behind liquefaction. Quake City also looks at what keeps buildings up, what knocks them down, and what causes landslides and avalanches.

The Anglican cathedral, Christchurch, New Zealand, early 2020, when restoration was about to begin.

One of the other Christchurch buildings, still awaiting restoration.

Earthquake waves.

Once upon a time, the waves were just mysterious shakes of unknown origin. People who lived near volcanoes saw that a few earthquakes were associated with volcanic activity, but the correlation is often hard to spot. For example, there are no volcanoes in Turkey, which is seismically active, though there are plenty of volcanoes dotted around other parts of the Mediterranean. It was easier to propose that a giant or a dragon under the Earth was stirring uneasily in its sleep.

Compare this with the truth, which is that our safe bits of solid ground, dubbed *terra firma* by the Romans, were in fact far from firm. In reality, most of the bits we live on are moving at a sedate pace, about as fast as a fingernail grows. They float on what is further down, and they move. Dragons are more likely…

Where the plates rub past each other, they grip and stick, they flex and tense up, and in the end, they slip again. Think of two small children, tugging on a rubber band: when it breaks, somebody is going to be hurt, as the energy stored in the stretched rubber is forced to relocate.

That energy is what we call an earthquake, but there is worse to come. All good things come to an end, and at some point, the plate that moves forward has nowhere to go. When that happens, it must plunge down, carrying all sorts of silt, sediment and water with it. An instability has been introduced into the Earth, and sooner or later, there will be a price to be paid.

These regions are subduction zones, the places where the world's largest and most dangerous earthquakes take place. The main subduction zones are in Japan, Alaska, Mexico, Central America, Peru, and Chile, all recognisable as earthquake centres. Of course, to some people, earthquakes are remarkably enlightening, thanks to the waves they generate, and the speeds at which they travel.

When you push one end of a Slinky spring that is hanging from threads, you can see a compression wave move along it. If you stretch a long rope and flip the end of it, you can see a wave of up-and-down motion travel along it. You can make a whole career out of studying the various waves that come from an earthquake, but for this discussion, I will stay with those two types. There is no mention of Love waves or Rayleigh waves here. (I didn't forget, I chose not to.)

Seismic waves travel quite a few kilometres each second. The actual speed depends on the type of rock, because like sound waves, seismic waves are affected by density. The speed also depends on the type of wave. The first waves to arrive are called primary waves, while the later ones are, as you might expect, called secondary waves. This is the simple version.

Given that, imagine that two cars have started together from another town. One travels at 50 km/h, the other at 100 km/h. If the slow car passes your house one hour after the fast car, how far away from you did they start? I leave it to you to work out why the answer is 100 km.

In the range 50 to 500 km away, P waves travel at about 8 km/s, while S waves travel at 3.45 km/s. The mathematics is a bit more complicated, but we can draw a circle marking all the places at the right distance. If we add several more circles from other stations, their intersection is where the quake happened.

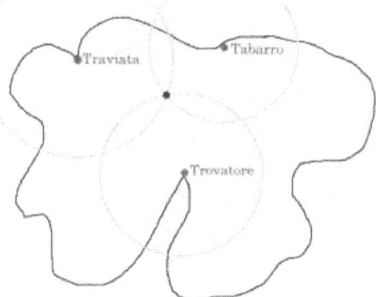

On the island of Trivia, three towns measured their distance to the focus, and so could locate it

Some of the compressional waves travel through the centre of the Earth, passing through rocks of different density, they are refracted, and with clever mathematics, we can learn about the innards of the planet. Maybe earthquakes aren't so bad after all!

Deep earthquakes.

Shallow earthquakes have their focus somewhere between the surface and 60 km down, intermediate earthquakes have a focus between 60 and 300 km below the surface, and anything more than that is called a deep earthquake. Deep earthquakes can happen as much as 600 km below the surface.

These only occur in subduction zones, where an oceanic plate is being pushed under another plate, down into the mantle. These earthquakes are a challenge to accepted wisdom, because below about 100 km, the temperature and pressure are thought to be too great for rocks to behave as solids. Instead, they flow rather like thick plasticine (modelling clay to American readers).

The cracks that form in stressed rocks near the surface cannot form when the rocks 'flow', and this means there should be no earthquakes happening anywhere deeper than 100 km. The standard explanation is that, below 400 km, the mineral olivine is likely to change into a different and denser form, called spinel. But sometimes, according to this theory, the olivine does not change straight away, and it plunges deeper and deeper, until the shear forces make it 'flip', producing the energy release that we call an earthquake.

Scientists (including me, earlier) are fond of quoting T. H. Huxley, who said *The great tragedy of science-the slaying of a beautiful hypothesis by an ugly fact*, and certainly that is the case with this hypothesis. A single deep earthquake in Bolivia in 1994, 640 km below the surface, seems to have given the old explanation a death blow.

The Bolivian area just happened to be covered with two temporary seismic recording systems, so the earthquake was accurately mapped, and it was fairly obvious the aftershocks were spread over a much larger area than you would expect from the olivine-spinel model.

This is not my field, and the last I heard, scientists were trying a number of explanations. Maybe the olivine-spinel model works to get the earthquake started, and then some other type of earthquake is triggered by that. Maybe the plate that is plunging down under Bolivia is kinked, and behaves oddly.

Or maybe there is some other system working altogether. All we can say for certain is that more evidence will probably lead us to a better understanding of what is happening, deep down in the earth, far from where any scientist can go.

Earthquakes away from the edges of plates.

John Bradbury, a Fellow of the Linnean Society, published his *Travels in the Interior of America* in 1817, and I came across him while researching a history of gold rushes, because I wanted the origin of two words, 'diggers' and 'diggings'. That led me to Bradbury's account of the lead mines of St. Genevieve, where the diggers did their digging with wooden shovels.

Bradbury was on the Mississippi on December 15, 1811 when the first of seven earthquakes that shook the USA along the New Madrid fault during 1811 and 1812:

> …I was awakened by the most tremendous noise, accompanied by an agitation of the boat so violent, that it appeared in danger of upsetting…I could distinctly see the river as if agitated by a storm; and although the noise was inconceivably loud and terrific, I could distinctly hear the crash of falling trees, and the screaming of the wild fowl on the river, but found that the boat was still safe at her moorings.…At day-light we had counted twenty-seven shocks.
> —John Bradbury, *Travels in the Interior of America*, 1817, 199–200.

This event was part of a series of earthquakes, one of them so massive that it rang church bells in Boston, 1700 km away, and caused part of the Mississippi River to run backwards for a short while.

There has since been little movement in the New Madrid Fault, especially compared with the San Andreas Fault, which has moved over 200 kilometres in the past five million years, so the chance of another earthquake around the

New Madrid fault is unlikely, but how many other unknown rifts might there be like this under other continents?

The tremors lasted so long and were so large that many of the frontier folk felt it was the end of the world, giving it not only a geological importance, but a theological importance as well. Across 1811 to 1812, membership in the Methodist church increased by 50% in the earthquake zone, compared with 1% for the rest of the nation.

The likely explanation is that during the last ice age, a huge ice sheet invaded North America, weighing down the crust of the continent. When the glaciers melted, freed from the heavy pressure of the ice sheet, North America slowly rose up. This glacial rebound continues even today and triggers tremors in Missouri, Kentucky, Arkansas and Tennessee.

So what lies beneath the earthquakes, and could they happen again? The question of a repeat performance is an important one, for what was mere backwoods in 1812 is now the world's most industrialized nation, and a quake strong enough to ring church bells in Boston, more than 1700 kilometres away would do huge damage now.

The 19th century quakes were felt in 27 of the states of the US, but about all they did back then was to topple trees, upturn flatboats and make the Mississippi River run backwards for a bit. The USA will not get off that lightly next time.

The rebound after glaciation is well-documented, and the crust is still lifting in Norway, Sweden and Canada's Hudson Bay. The bad news is that big earthquakes may continue along the New Madrid fault for the next 10,000 years, long enough for North America to bounce back fully from the glaciers. And the really bad news is that these events seem to happen every 200 to 900 years—and the 200 years is now up.

Australia has had very few serious earthquakes because the Australian continental plate is just too old and too thick. As a result, when Australia moves north, the island of Papua New Guinea acts like the crumple section at the front of a car, grinding, breaking and doing all of the suffering, while the passengers behind get off without so much as a scrape.

But that need not always be the case, for sometimes continents can split apart, forming a rift valley. America's New Madrid fault lies in the centre of a large continental plate, and contradicts the usual pattern of earthquakes happening where plates collide. Is North America coming apart? Probably not…

A short history of seismology.

This is the science of measuring and describing earthquakes. Today, much of the work is done using delicate instruments, but the earliest known instrument, constructed by a scholar named Zhang Heng in 132 CE, was rather more robust.

It was about two metres in diameter, and consisted of a large vase with a set of eight bronze dragons holding delicately balanced balls in their mouths. When an earthquake struck, one or more of the balls would fall from the dragons' mouths into the mouths of bronze frogs, placed below the dragons, indicating a direction towards the epicentre.

A 1953 Chinese postage stamp of the seismometer.

The secret of Zhang Hang's apparatus was a heavy weight, hanging in the centre of the vase, and connected by levers to the dragons' mouths. As the ground swayed and the outside of the vase swayed with it, the heavy weight moved rather less, due to inertia. The levers caused the dragons' mouth on one or more dragons to open, dropping its ball to a waiting frog.

The first modern scientist to think carefully about earthquakes was John Michell. In 1760, five years after an earthquake that almost destroyed Lisbon, Michell suggested that earthquakes were waves passing through the earth, and that an accurate timing of the arrival of the waves could be used to locate the centre of an earthquake. Michell also related earthquakes to volcanoes, and suggested that the Lisbon earthquake started under the sea floor.

Luigi Palmieri (1807–1896) was an Italian physicist who made a sort of seismograph, using horizontal tubes, partly filled with mercury, but these required continual watching, which would be boring. Since Palmieri had his apparatus set up on the slopes of Mount Vesuvius, boredom was less of a problem. John Milne was an English geologist who lived in Japan for 20 years from 1875, so he also had plenty of opportunities to study earthquakes. In 1880, he invented the pendulum seismograph that we use today.

Basically, the seismograph relies on the fact that a large hanging mass will not move fast in any direction when the ground underneath it vibrates rapidly.

So if a mass like that has a pen attached, and the pen can leave a trace on a moving piece of paper that is attached very firmly to the ground, the paper will vibrate, the pen will stay still, and a trace will be left behind.

Tsunamis.

> The name *tsunami* comes from the Japanese language, and means "harbour wave", because boats out to sea will generally feel and see no wave at all, but close to the shore and in harbours, all the energy of an 800 km/h wave is used to form a much higher but rather slower wave. Sometimes, a harbour focuses the power of a tsunami.
>
> But why use a Japanese word? Japan is largely volcanic, sitting on a subduction zone, and has long been a literate society. The Japanese people were, and are, aware of tsunamis, all too painfully aware. It's a good word, while "tidal wave" is misleading, so tsunami it is!

A tsunami is a water wave generated by sudden earth movements. Tsunamis may travel thousands of kilometres as barely visible waves before hitting a coast. In shallow waters, a tsunami builds up to a considerable height, and may flood a large coastal area, without any warning, far from any seismic activity.

Tsunamis are sometimes as high as a five-storey building when they reach shallow water. They travel at about 1000 km/h in the open ocean and at around 150 km/h closer to the shore, so if they are identified near their sources, which may be on the other side of the Pacific ocean, people could get to safe places before the tsunami arrives, saving many lives.

People called tsunamis 'tidal waves' until 1994, when a disastrous quake on December 26 killed more than 300,000 people in those parts of Asia facing the Indian Ocean. It was the first major tsunami of the Media Age, and by the time its carnage had become old news, the world had acquired a new word.

In hindsight, the risk of a tsunami was obvious. A huge earthquake had happened, south of Australia. This indicated a major shift in the crust of the Earth, which caused scarcely a ripple because the plates moved sideways, but that movement transferred tension from way south of Australia to the north, where the north-pushing Australian plate would be more likely to cause a vertical movement and trigger chaos.

Sooner or later, the Indian Plate, pushed by the Australian plate, had to slide under the Burma micro-plate, bumping it upwards, like this:

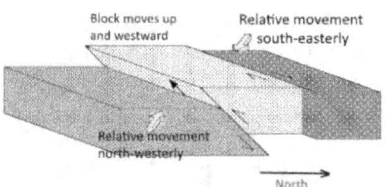

With the southern (red) plate (the Indian plate) moving north-west, relative to the northern (blue) plate (the Burma micro-plate), a block (light blue) broke away, and it was pushed to the west and upwards.

In the 2004 event, a 1200 km rupture opened lengthwise to the north-northwest at 2.5 km/s in the first 10 minutes of the earthquake. More than 30 cubic kilometres of water were displaced by the shifting sea floor, and this generated the waves that did the damage. At sea, they were about 1-metre high, but up to 15 metres high as they came ashore. With no warning, people died in 14 different countries.

Looking back, the risk was obvious, but giant tsunamis are rare enough for people not to think about them. They are caused when something large like an asteroid or a cliff falls into the sea, when a major slide of material happens under the sea, or when the ocean floor heaves after the crust breaks.

The Atlantic coasts of Europe, Africa and the Americas are at risk from the Canary Islands, where a future eruption of the Cumbre Vieja volcano could trigger an undersea collapse, with a block of rock "…twice the size of the Isle of Man…" falling at up to 350 km/h. The energy released would be equal to the electricity used in the USA in six months, and it would mostly go into the water.

The first result would be a dome of water, 900 metres high and tens of kilometres wide, generating a series of crests and troughs, surging out at 800 km/h. The waves would spread, losing energy with distance, so the 100 metre waves on the western Sahara shore, would be 50 metre waves to Florida and the Caribbean, 8 to 9 hours later. The waves hitting Brazil could reach 40 metres, but in some places, the waves could be funnelled, as they were at Hilo in Hawaii in 1960. The only real certainty is the speed of transmission.

Sooner or later, an asteroid will plunge into an ocean. Asteroid 1950 DA is one kilometre across, and there is a slight chance of it hitting us in 2880, but there are probably other threats, still to be spotted. An asteroid that size would blast a cavity about 18 km across, all the way down to the sea floor. As the ocean rushes in, waves will radiate out at 800 km/h, and rather than the one big wave that movie-makers like to show, there will be a number of them, starting small and getting larger, one every three or four minutes.

This variation is important if you are ever caught in a tsunami zone: in Sri Lanka in 2004, the biggest crest was the third or fourth, which gave a British geologist a chance to warn staff and tourists to clear the beach. In the ocean, those waves were a metre high, but their tremendous speed converted to extra height in shallow waters. The wave system can also lead first to a 'drawing-down', where the sea appears to go out: if you see this happening, run!

When in doubt, hurry up the nearest hill!

In 1998, a landslide caused by a magnitude 7.0 earthquake triggered a tsunami on the north coast of New Guinea, sending water surging in at 10 to 20 m/s, about 30 to 60 km/h. A wind at the top end of that scale can buffet you, but the force of a water wave is about a thousand times as high as wind at the same speed.

Making gravel on Hawaii.

In the year 2000, I was working as an in-house science writer, and a story came my way about a tsunami that never was. How could I resist? The Hawaiian islands of Lana'i and Moloka'i have gravels high on their southern coastal slopes, and these have been interpreted in the past as leftovers from a giant tsunami.

Then late in 2000, uranium-thorium dating revealed an entirely different picture. The study looked at the Hulopoe gravel on Lana'i, an extinct and eroded shield volcano that is about 1 million years old. The lower slopes contain large amounts of bioclastic gravel, which is gravel generated from living material, mainly coral in this case, with concentrations in gullies at about 70 metres above sea level, and in isolated outcrops at about 170 metres above sea level.

The dating study showed that the gravel contains coral of two different ages, about 135,000 years and 240,000 years, with significant geographical and stratigraphic ordering. So the gravel was formed as multiple deposits, separated by large time gaps, ruling out the single-giant-wave version.

Instead, it seems as though the gravels were simply deposited in a normal way at a time when sea levels were high, and then reached their present height by the later uplift of Lana'i. The finding has relevance well beyond Hawaii, though.

A Lana'i tsunami about 105,000 years ago is supposed to have shaped the east coast of Australia. But if Lana'i itself was unaffected, it leaves open the question of what caused the apparent tsunami damage in Australia.

In an interview with Ken Rubin, one of the researchers, I mentioned a typo in the paper where the east coast of Australia was wrongly labelled as the west coast, the sort of thing that any busy writer does from time to time. He agreed that it would have been nice if the slip-up had been detected in the peer-review process, but said that his real concern was that the peer review process seemed to be unable to kill off the enthusiasm for scenarios which involve exciting catastrophes.

He believes that even though their evidence has shown that there really *was* no tsunami, once the catastrophe story is out, it will probably never stop running. He is probably right, but for what it is worth, the Lana'i tsunami never happened—or if it did, it left absolutely no traces on Lana'i. Which version would you prefer?

8: Rocks that ignore the rules.

Tilted, folded beds near Todra Gorge, Morocco.

Floating pumice.

Throw a rock into water, and it splashes and sinks, unless it's pumice. Think of pumice as a frothy glass, a gas-filled rock full of bubbles and tubes, a rock produced when gas fizzes out of extruding magma. The bubbles give it a low density, so the pumice usually floats in water.

Volcanic glass forms naturally when magma cools so fast that the atoms don't have time to connect in regular lattice structures and crystals. The magma involved is silica-rich, making a sticky, thick lava that froths up, a bit like the head on beer. It cools fast when it is extruded into water, locking the bubbly structure in place.

Marine life on pumice from the 2012 L'Havre Seamount eruption: bryozoan on the left, tubeworms on the right. The scale is an Australian $2 coin, 20 mm across.

Until recently, people thought pumice only came from volcanic eruptions in shallow water. The 2012 L'Havre Seamount eruption (mentioned in the preamble) showed that was wrong, because this source was a deep submarine volcano. That eruption produced the two inhabited pumice pieces shown above, the fragments that inspired this book.

"Lava Soap" contains crushed pumice, and when you wash yourself with it, the effect is the same as wiping crushed glass over your skin: you cut off the outermost layer of skin, and any dirt engrained into it. (A quick safety note: you *can* make pumice dust by rubbing two pieces together, but this is *very dangerous*!)

Floating plates.

When I was a beginning science teacher, I used to like stirring young minds by exposing them to paradoxes. One of my toys was a large measuring cylinder with a layer of mercury, under a layer of water, under a layer of oil.

Into this, I had inserted a small rock, a small block of hardwood and a small cork. The rock floated on the mercury, the wood was supported by the water, and the cork lay on the top of the oil. There was no explanation: it just sat in a glass case in a public passageway.

Students either worked it out for themselves, or they found out which teacher changed the items in the display case (me) and asked questions, or they asked other students who knew the answer. My early answers were more of the nature of hints like "Why did Archimedes run naked down the street, shouting '*Eureka*'?", though I made it easier later.

No, I won't explain it better than that right now, but think of yourself floating in a bath: you sink down into the water until the amount of water you displace has the same mass as you. If a gangster has fitted you with concrete-filled boots, the amount of water you displace may be less than your mass, so you will sink to the bottom.

Of course, if you have a suitable flotation device attached, you may float, even with the concrete boots. This holds in every case where something floats in a fluid, if it is partly submerged, whether it is you, an ocean liner, an iceberg or a mountain range. All floating things displace their own mass of fluid, so given the right conditions, even rocks can float on something.

This effect was first noticed when 19th century surveyors were mapping parts of India, about 100 km from the Himalayas. Surveyors used to depend mainly on the simple principle that the angles of a triangle add up to 180° plus the firm belief that a plumb line points directly at the Earth's centre.

That opinion rests on an understanding of gravity (chapter 13): all mass exerts a gravitational pull, and the closer that mass is, the stronger the force is. So the Himalayan mountain range ought to pull the surveyor's plumb line ever so slightly sideways. But how would you know that this was so?

The answer is simple: you can determine where true vertical is, by taking sightings on the stars. When surveyors did this, they found that the difference between the plumb line and the true vertical was 5.236 seconds of arc. There was just one problem: careful calculation showed that the discrepancy ought to have been 15.886 seconds, just over three times as great. When the numbers are checked and found to be right, that means some of the assumptions behind the calculations have to be wrong.

The surface rocks, the crust, have a specific gravity of between 2.75 and 2.90. The underlying rock must be more dense, because the total mass of the Earth is 5.97219×10^{24} kg which gives us an average specific gravity for the planet of 5.513. Those are numbers, and they cannot be argued with. We can, however, question the assumption that the Himalayas are a lump of less dense rock, sitting on top of more dense rock, like a cork on water.

Even a cork sinks into the water a bit, and the planet's mountain ranges float on the more dense rocks of the crust, which have s.g. values in the range 2.90 to 4.75, so the mountains sink into the denser rocks. We cannot easily dig or drill down to examine the roots of a mountain, but cunning observation of seismic waves allows us to 'see' what is down there.

How plate tectonics works.

Originally, 'tectonics' applied only to movements on a fairly small scale ('small' as geologists see things—about the size of the Swiss Alps). Many mountain chains show evidence of tilting, folding, and faulting in rocks, and these seemed to explain a lot of geological features, especially those confusing cases where marine fossils appear on mountain tops.

During the early 1960s, the notion of 'continental drift', was replaced with the far more acceptable plate tectonics. This idea differed from continental drift because it offered a way of explaining why the continents were 'drifting' (they weren't drifting at all—they were being pushed along by convection currents).

There was now a way of explaining huge features like the Himalayas, central America, and even Papua New Guinea. This was because we could now see that really large pieces of the crust, like the Indian sub-continent, were being pushed into the rest of Asia, and forcing the land at the join to rise up into the Himalayas.

How Africa and South America 'fit'.

You only had to look at a map of the Atlantic Ocean to see how Africa seems to fit in neatly against South America. In 1596, a Dutch map maker called Abraham Ortelius (1527–1598) suggested that the two sides of the Atlantic had been torn apart, but he did not suggest what might have done it.

The geology on opposite sides of the conjectured break also matched fairly well. Once biology got going and people started collecting plants and animals, interesting parallels showed up, like the presence of monkeys on both sides. You could explain Asian monkeys by assuming they had wandered eastwards from Africa (or *vice versa*), but the South American monkeys were a puzzle.

A close inspection showed that these New World (South American) monkeys were quite different from the African ones, suggesting a great deal of evolution had happened since the two monkey groups separated. Other plant and animal distributions made more sense if continents had originally been joined together.

Then there were things like marsupials in both Australia and South America and similar fossils on both sides, such as platypus teeth found in South America. Platypuses or platypodes (*never* platypi: trust me, I'm a pedant!) have no teeth these days, except in juvenile animals, but an early Cretaceous platypus, *Sterodon galmani*, had teeth, and in 1992, the teeth of a South American platypus, *Monotrematum sudamericanum*, were found, dated to 61 million years ago. The secret to these is that in the past, there was a link between South America and Australia through what is now Antarctica.

In 1912, a German meteorologist named Alfred Lothar Wegener (1880–1930) published an account of how continental drift might have happened. He thought the supercontinent Pangaea began to split, about 200 million years ago.

There were two problems: there was no way to explain how the continents were being moved around, and if the continents and oceans were getting larger, as some people thought, there was no way to explain where the extra space was coming from.

One of the wilder explanations was that the earth was getting larger, as it gathered dust and meteorites from space. The key find turned out to be the distribution of a fossil fern named *Glossopteris*, found in South America, southern Africa, Australia—and Antarctica. It took people a while to realise that the *Glossopteris* fossils were the key.

The revolution began with the idea of sea floor spreading, which came from mapping the sea floor, at first carried out with long weighted lines, lowered to the floor, and later with sonar: sending ultrasonic 'pings' at the sea floor and timing their return. This revealed the shape of the seabed.

The first chart showing parts of the mid-Atlantic ridge appeared in 1855. Ships laying cables across the Atlantic also detected parts of it, then in 1947, cores of the seabed showed that the sediment on the floor of the Atlantic was much thinner than it should have been under an ocean that had existed for 4 billion years.

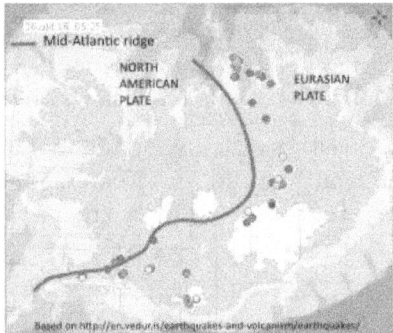

The location of Icelandic earthquakes in a short period in 2016, with the position of the rift added in.

The answer we accept now is that there was a rift, a ripping-open of the world, running up the Atlantic, and passing through Iceland. Soon, more ships were mapping sea floors and tracing the whole of the global mid-ocean ridge, more than 50,000 kilometres long and, sometimes more than 800 km across. The ridge was no mere range of hills, either, but a range of underwater mountains that averaged 4500 metres above the sea floor.

This is the view from the bridge between the two plates.

The exposed rock on the American side of the rift, and again in a closer view. Can you see how lava flows have run over each other?

The sea-floor basalt had a peculiarity, in that the magnetic fields sometimes go the 'wrong' way, the reverse of today's magnetic field. We know now that every so often, there is a polar reversal, where the Earth's magnetic field 'flips', reversing the magnetic north and south poles.

As liquid basalt escapes from deep in the Earth and turns solid, the magnetic field of the moment is recorded in the rocks, so the basalt has a sort of date stamp on it, but there is a trap to watch out for. On maps of the zones of normal and reversed magnetic fields around the mid-Atlantic ridge, you see a pattern of stripes going across the sea floor.

The way this is shown in books, people often think the sea floor is striped like a zebra or a bar code. It isn't: black and white in the maps are there just to show the two polarities, and all the basalt is dark, almost black. The 'stripes' are of different sizes, reflecting longer and shorter periods between reversals, and the two sides of the ridge show a mirror pattern.

By 1961, people were beginning to hint, rather nervously, that maybe the basalt was oozing from the floor and spreading out to either side. During the 1960s, deep-sea drilling rigs began to bring up cores from the sea floor, and by 1968, fossil and isotope tests on the cores established the sea-floor-spreading hypothesis, with the youngest rocks near the ridge, and the oldest rocks further out. Here is the evidence that made plate tectonics come alive added to this original 'continental drift' evidence:

• sea floor spreading on the oceanic ridges and 'magnetic striping' of the newly-formed rocks, caused by polar reversals;

• evidence of subduction zones where rock was being forced down inside the earth; and

• measures of continental movement which were now able to be made.

This sign is on a bridge across the mid-Atlantic rift in Iceland.

Once this theory was worked out, mainly between 1963 and 1966, and once the evidence had been collected, we could explain all of the observations, and be left with no puzzling problems, so long as we accepted that the small-scale tectonics of geology had a big brother: plate tectonics.

Thingvellir in Iceland is the site of the oldest parliament in the world, and another place where we can look at the newest parts of the planet's surface.

Now we can account for the more peculiar earthquake areas. Spreading in one place means rocks being buried somewhere else. The subduction zones where one plate slides under another, the deep sea trenches, the position of Wallace's Line, and even the origins of Africa's Rift Valley, where many of the earliest human and pre-human fossils are found today, were all explained.

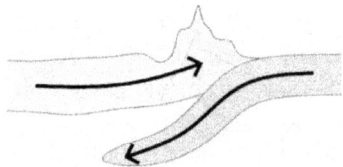
The blue plate on the right is plunging under the yellow plate, forming mountains on the yellow plate.

Plate tectonics explains the continent shapes and plant and animal distributions, but it also explains the main mountain regions like the Himalayas and the Alps, the distribution of volcanoes and earthquakes, the location of island groups like Hawaii and the Aleutian islands—and the forces that drive the process.

The Himalayas, the Swiss Alps, and the Andes are all formed as the crust piles up where plates are colliding. The volcano-free earthquake zones of Turkey and Greece are explained: the movement between the plates there is not the sort that generates volcanoes. Around the Pacific, the Ring of Fire, the long chain of active volcanoes is explained, while the Hawaiian islands are the result of a plate slipping over a 'hot spot' that keeps generating volcanoes.

It is fairly rare, but sometimes, pieces of sea floor may be lifted up above the sea surface. This sort of formation is called an ophiolite and it reveals hidden processes.

The oldest evidence for plate tectonics found so far lies in a piece of ocean crust found in 2000 in a mountain belt in the Eastern Hebei Province, just a short walk from the Great Wall of China. To be precise, it *was* oceanic crust, once, back when the world was young, some 2505 ± 2.2-million years ago, based on the uranium-lead ratio in zircons. That's on *my* bucket list!

In summary, the key to plate tectonics is that continents are made of less dense crust rock that floats on the more dense mantle rock, and the crust forms plates that can be pushed around. This continental crust makes up the continents, although small parts of the land, like part of central America, are made up of material that has been uplifted by tectonic forces.

The moving plates, as they collide, cause massive upheavals in the form of earthquakes and volcanoes. Areas of mountain lie above very thick crust, in just the same way that large icebergs extend further above and below sea level. The surface of the planet is shaped by plate tectonics.

We can measure the actual movement of tectonic plates today by GPS stations located at fixed points. On the evidence, this slow movement ripped apart a supercontinent that is now referred to as Pangaea. It later divided to parts now given the names Laurasia and Gondwana. The southern hemisphere distribution of some groups of plants and animals reflects their origins in Gondwana. Laurasia gave rise to most of the northern continents.

A plate tectonics timeline.

In 1595, Abraham Ortelius suggested the Americas were "torn away from Europe and Africa by earthquakes and floods", an early version of continental drift.

In 1620 Francis Bacon pointed out the jigsaw fit of the opposite shores of the Atlantic Ocean, a faltering step to drifting continents and plate tectonics.

In 1910, Alfred Wegener noticed the same close 'fit' between the west coast of Africa and the east coast of South America and started thinking about continental drift.

In 1912, Wegener elaborated his theory of continental drift based on fossil and glacial evidence, and first spoke of continental drift.

In 1926, geologist Arthur Holmes said the Earth's internal heat had to go somewhere, and argued that there may be convection currents in the Earth.

In 1960, Harry Hess proposed that new sea floor might be created at mid-ocean rifts and destroyed at deep sea trenches, a key to plate tectonics.

In 1963, Vine and Matthews explained the 'stripes' of magnetized rocks as due to sea floor spreading and the periodic geomagnetic field reversals.

> Note: this page was written while listening to *The Alfred Wegener Song* by The Amoeba People, on Youtube. The rest of the book ran on Sibelius.

9: Water and geology.

Iguazu Falls, Brazil.

There are two useful kinds of water in the ground. Artesian water is trapped between two layers of rock that are waterproof, so the water pressure is enough to make water gush out of a well. Sub-artesian water is similar, but the water needs to be pumped to the surface. The second sort is groundwater.

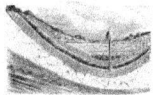

An artesian well diagram based on one in a 19th century issue of *Scientific American*. The well taps the grey aquifer, which is between the black impermeable bed above and the yellow impermeable bed below.

In the early 1860s, there was a race on in Australia, with different colonial capitals desperately seeking a route that would bring the telegraph line from Asia (and so from Europe) to their city first, because having knowledge first was power.

Famously, several members of the ill-fated 'Burke and Wills expedition' (including both Burke and Wills) died of starvation, scurvy, thirst and stress. Far to the west, a canny Scot, John McDouall Stuart, made repeated assaults through central Australia and eventually succeeded in finding a route. More importantly, it was a route which had water, in the heart of a brutally dry land.

By a combination of good luck and good management, Stuart worked his way along the edge of the Great Artesian Basin, a monster underground lake that lies beneath 22% of Australia. It draws water in, way off in the lush tropics of coastal Queensland, so the water can seep through porous rocks that are protected from the drying sun by a massive overburden of rocks, sand and dust.

At the end of its journey, the water emerges, carrying dissolved salts that crystallise out to form characteristic piles called mound springs. It was these springs which made Stuart's telegraph route work, because they gave water to the operators in the repeater stations, dotted through central Australia.

The water in a good aquifer may flow 125 metres a day under the best conditions, 15 metres a day under average conditions. Because the flows occur in such a broad 'pipe' in the Great Artesian Basin, there is little need to worry about how fast the water flows: the big worry is how fast it is being lost. Only in recent years have people realised that the losses are happening too fast.

The early Australian settlers came from a land where water was often drawn up from wells. They knew, without understanding why, that there was water beneath the ground, and when there was no surface water, they sank wells, looking for what we now call groundwater.

Rain soaks into the soil, although in heavy rain, water also runs into creeks and rivers. The part of the water in the soil that isn't captured by plants keeps sinking slowly, pulled by gravity, until it reaches a zone where it can go no deeper, and then it fills in all the gaps in the soil, and begins flowing sideways across the bedrock.

Under every dry creek and river, there is saturated soil, and there are slow-flowing underground creeks and rivers, as the original Australians knew and showed the settlers, but this is water that was rain not long before.

Groundwater occurs wherever the geology allows it to exist, provided there are surface supplies available to top it up as it seeps down to the sea, or is taken out. The right geology means having rocks and sediments that are either porous or permeable, and in geology, these two words have different meanings.

A porous rock is one with spaces where water can fit, a permeable rock is one where the spaces are linked together, so water can pass along through the rock. To hold water, a rock or a sediment layer needs pores that can hold the water, but being porous is not enough: pumice is full of pores, but the pores are not connected, and so the rock is not permeable.

Gravels and sandstones are usually very permeable, letting water pass freely through them. Water falling as rain may either be absorbed by plant roots, or sink down until it can sink no further, either because the rock below is impermeable, or because it is saturated with water.

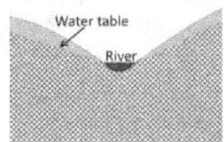

A simplified groundwater diagram.

The water spreads out, forming a layer known as the water table, but despite its name, the water table is rarely flat and level. The ground below the water table is saturated with groundwater which flows slowly to the sea or lakes. Wells fill to the level of the water table. Groundwater moves sideways through an aquifer with a rate of flow that depends on the aquifer's permeability and the slope it is flowing down.

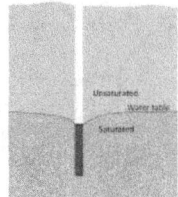

The water table is lower near a well.

When water is drawn from a well, this lowers the water table in the immediate area, and more water flows in. Over time, with continued pumping, the local water table takes up a conical shape. The slope of the cone gets steeper when water is drawn from a well faster than it is replaced by inflow or local recharging.

We humans have three main stores of fresh water. The first is on the surface in rivers, lakes and dams. The second is water locked up in the ice of glaciers, and the third is water in the ground. Depending on the nature of the ground and its rocks, some parts of the world have more of their water flow under the ground than on the surface.

At a soak or a spring, the water table reaches the surface; at a river or a lake, the water table also reaches the surface, instead of being somewhere below it.

Wet sand dunes and rocks behind a beach near The Entrance, NSW, with emerging groundwater.

The water table eventually feeds into either a river or the sea, lowering the local level, but more water flows in. The key points to understanding groundwater: it flows beneath the ground in a permeable aquifer, but it flows more slowly than water above the ground.

In Standley Chasm in arid central Australia, water comes to the surface, close to rock.

Occasionally, the water is trapped beneath an impermeable layer, so the pressure builds up, and the *potentiometric surface*, the level to which it could rise is higher than the ground. When this happens, a well drilled into the aquifer lets water flow out under pressure, forming an artesian well.

> One of the most wasteful uses of artesian water happened at Thargomindah in 1898. This tiny town, about 300 km east of where Burke and Wills died, lies in dry country, and water had to be hauled there from the Darling River at Bourke in old ship's water tanks.
>
> Then a bore found water at a depth of 800 metres. Around 1894, a Thargomindah sawmill owner set up a steam-powered generator, and the local council bought this in 1898. Two tenders were submitted to operate the generator, the cheaper quote coming from a local blacksmith, who made a water wheel, mounting this in a casing made from an old water tank.
>
> A water wheel may seem unlikely in flat dry country, but Thargomindah's bore delivered hot water under such high pressure that uncontrolled, the water spouted more than 20 metres high, and this pressure was enough to spin the wheel and generate electricity!

The best estimate of the earth's groundwater is about 50 million cubic kilometres, but we could only reasonably hope to extract about 4 million cubic kilometres. This leaves out the deep groundwater that we can probably never use, and some very salty water that is also found below the surface.

A consciousness of streams.

I penned this essay while writing this book, and set it aside, before I decided that, with the disclosure finishing the box below, it still has a place here.

> There is an integral link between beer and geology, and it dates back well into the 19th century, when it was considered appropriate for ladies to attend geological outings. Note the word "ladies": we are talking middle to upper class adult females here, people who were supposedly complete wowsers, yet at the end of each excursion, they would all pile into a pub to drink beer. (Obligatory disclosure: The author is an active member of the Facebook group **Geologists Drinking Beer**.)

Many tales of adventure have at their centre a group of very different companions who join together for a quest or a journey. In Tolkien's books, hobbits, dwarves, elves, wizards and humans rode together, The Magnificent Seven went in another direction. Tripitaka, Sandy, Piggsy and Monkey did things one way, and the Guardians of the Galaxy took yet another path.

Sometimes the travellers separate, only to join up again later, though Chaucer's Canterbury pilgrims mostly stayed together. The pilgrims were all English, so they were more homogeneous. I prefer to liken a river's contents to Frodo's band, because like Frodo's friends, the things you find in a river are likely to wander, even as they head in the same direction.

Let's start with the water molecules: some of them evaporate, go up and become clouds, some of them are taken up by algae that get eaten in turn by a tadpole, a fish and a heron, before being excreted back into the water to continue their journey. The clouds may return as rain.

Some of the water molecules seep into the ground that lies beneath the river, to become groundwater. Now they make their way to the sea a little more slowly, but one day, the water molecules will all get there. Rivers also contain dissolved salts, and that term needs explaining. Throw a piece of sodium metal into the river, and it will fizz around on the surface, reacting with the water and making hydrogen.

Use a big enough piece of the metal, and the hydrogen will explode, sending hot sodium all over the place (I know: I used to be a chemistry teacher). Chlorine is a poisonous green gas that makes the water acidic, but dissolved sodium chloride is harmless enough, in small amounts.

Matter is made up of atoms, and a good rule of thumb is that the chemical properties of an atom or group of atoms will depend on:

* the size of the lump;
* the shape of the lump;
* the charge on the lump;

* the preferred charge of the lump; and
* the distribution of any charge across the lump.

Metallic sodium and gaseous chlorine are both uncharged: their reactions with water happen because sodium is more stable if it gets rid of an electron, and chlorine is more placid when it gains an electron. The sodium and chlorine in our river are there in placid form as fairly polite sodium ions and relatively gentle chloride ions, with positive and negative charges respectively.

The sodium ions have a good chance of making it all the way to the sea, without hesitations, delays or diversions, but they may also be taken up. The smaller number of calcium ions are less likely to make an uninterrupted journey, because plants need calcium to make cell walls, some invertebrates need calcium for their shells, and vertebrates need calcium for their bones and teeth.

We know far more about the residence times for ions in sea water, and I will come to that later, we do know that something odd happens, either on the way to the sea, or in the sea itself. Scientists all agree that the salt in the oceans came from rivers, but while ~85% of all oceanic ions are sodium or chloride, river salt is ~9% chlorine and 7% sodium.

River salt has 12% sulfate, 5% magnesium and 17% calcium, while silica is also more common (up to 10% of the total salt) in rivers than in the sea. By mass, sea salt is 55% chlorine, 30% sodium, 8% sulfate, 4% magnesium and 1% calcium. There are few helpful data in this area, but the 2011 Australian Drinking Water Guidelines offer the following limits, all expressed in milligrams per litre (ppm): unless otherwise indicated as "aesthetic", these are health limits.

Aluminium, 0.2 (aesthetic); barium, 2; beryllium, 0.06; boron, 4; cadmium, 0.002; chloride, 250 (aesthetic); iodide, 0.5); iron, 0.3 (aesthetic); lead, 0.1; mercury 0.001; nickel, 0.02; silver, 0.1); sodium, 180 (aesthetic); sulfate, 250 (aesthetic); zinc, 3 (aesthetic). As well, the river will contain carbonate, fluoride, nitrate, phosphate and other negative ions. The other thing that varies is the total mass of salts: in river water, values are typically 500 mg/L, up to 2000 mg/L, half a gram to two grams in a litre, while sea water contains around 35 grams in one litre. Together, these make up the Company of the River.

Iran's sideways wells, or *qanats*.

Almost 3,000 years ago (nobody is sure exactly when), a clever hydrologist in modern Iran invented the *qanat*, an upward-sloping tunnel that creeps in, just under the top of a sloping water table, to let water flow out under gravity alone.

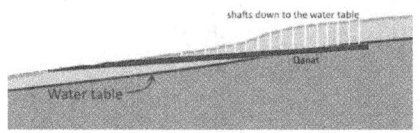

How a qanat burrows in under the water table.

As you move across a plain towards a mountain range, the ground rises, so too does the water table, producing a sloping surface which can be used. A qanat usually starts as a series of wells which allow the qanat makers to work out where the water table is, and whether it is worth while cutting a qanat at that point.

In the end, the qanat was made by digging shafts down to the water table, and then linking them by a sideways tunnel to let the water run down to the shaft below, always getting closer to the surface of the ground. At the end, the water flowed out into a basin or pond, where people could collect it.

The idea spread and, by 550 BCE, there was a qanat on the Greek island of Samos. With the spread of Islam in the 8th and 9th centuries, a single culture covered a far larger area, and the idea reached northern Africa, gaining new names as it went, so the Iranian qanat is known in Morocco as a *khettara*. These clever water sources can be seen today in most Muslim countries, and also in Japan, Italy and Luxembourg.

A *khettara*, the Moroccan name for a *qanat*.

Salination.

Groundwater supplies are a strictly limited resource, but they can be recharged in some areas by sensibly diverting stormwater into the aquifer. When too many trees are cut down, rainwater can reach the water table faster than it flows away, so the water table rises, sometimes reaching the surface. This brings dissolved salt with it, and this can kill plants. We call this process salination.

Salt water has become a serious problem in Australia in recent times, because the level of salty water has been rising, often all the way up, so the saline water table breaks through the surface, scalding the earth and killing all plant life.

In the past, the inland plains and hills of Australia were covered with bush, mixed forests with deep-rooted trees and shallow-rooted smaller plants. The water table was a long way under the surface of the ground. In an average year, it is likely that only about 0.1 mm of a 250 mm rainfall made it down to the water table: the rest was captured by roots, and taken back to the atmosphere.

Now, each year, between 3 and 20 mm of the annual rainfall slips past today's plants, and sinks down to add to the water table. This is too much water for it all to creep away, down to the rivers, and the water tables are rising.

The original small addition to the water table was balanced by the water that ran away into the rivers and down to the sea. Any rain that escaped the roots of the smaller plants was soaked up by deep tree roots, so the water table remained far below the ground, and the water soaking through to the water table was balanced by a slow trickle away, until the salt water ended up in the sea.

The larger trees with deep roots act as "biological pumps", taking water from deep underground. Mallee trees are a kind of low *Eucalyptus* shrub, with roots which go down as far as 17 metres below the surface. As the water evaporates from tree leaves, the surface of the water table is either lowered or kept down.

So long as there are mallee trees in an area, the water table is held to that level. So long as there are trees on the hills surrounding a plain, water will not elude their roots, to join the water table and flow down onto the plains below.

Around Uluru, a she-oak called *Allocasuarina decaisneana* ("desert oak") can have roots that go 10 metres underground, and almost any tree growing in dry country will have similarly deep roots.

Grassland dotted with desert oak, near Uluru.

With white settlement, much of the bush was cleared for pastures on the hills, or for crop fields on the plains. The crops would be entirely absent for parts of the year, and the pasture grasses were less resistant to drought, so more water was able to soak down past their shallow roots when rain came at the end of a drought. The trees had been chopped down or ring-barked to let more sun through to grow more grass, so any rain getting past the first few centimetres of the soil was free to trickle down until it found the salty water table, far below.

That is, the salt used to be far below. There is an increased flow from the hills, and sooner or later, the salty groundwater breaks the ground surface. When it does, the plants die, and the soil loses another defence.

The answer lies mainly in planting trees on the hills, trees that will drill deep for water, hauling it to the surface and stopping it reaching the plains. Given time, the water table will drop again, and given more time, rain will wash the surface salt back down into the depths, but it will all take time, and it need never have happened if people had understood the basics of hydrology.

Salination has happened many times in the past, in Africa, the south-west of the United States, across the Middle-East and elsewhere, but the Australian case is happening right now, although many of the necessary steps to cause the problem were taken fifty or a hundred years ago.

The origin of the Australian salt is a bit of a mystery. Once, people thought the salt was carried in when the sea level was higher, and shallow oceans covered much of Australia, which is a remarkably flat, low continent. You will still read this in some books, but scientists now think the salt was blown in as salt spray, at some time in the past.

When the water table is just one or two metres below the surface, rainwater is unable to carry salts down out of harm's way for the crops and other plants, and the trouble begins. First, some of the plants die, so even more water gets through, then as the water table comes from the depths back to the surface of the soil, it carries salt with it, and then deposits the salt on the surface. All plants need small amounts of minerals, but not that much!

The really productive ground is the fragile shell of rich topsoil on the surface of our planet, just centimetres thick in some places. If the plant cover is killed by salt, there is nothing to hold the topsoil. It blows away, and there will be no crops there for a long while. Bush killed by salt is a ruined wasteland where nothing thrives.

John Snow's cholera map.

This last portion of this book was written just after "the Wuhan coronavirus" became Covid19, but in early 2003, a coronavirus disease called Severe Acute Respiratory Syndrome (SARS) emerged in China. At first, the Chinese authorities ignored it, and soon SARS spread in a major way to Vietnam, Hong Kong and Canada, carried by air passengers who reached their destinations before any symptoms showed up. In those countries, it killed about 14–15% of those who were given good treatment.

Within a few weeks, virologists knew what caused the disease, and epidemiologists knew a lot about where it had come from, though even today, we are unsure what the original source animal was. The disease was stopped because epidemiologists knew enough about the cause.

Epidemiologists are medically trained but to use knowledge, statistics and logic. If children living on a river get bad diarrhoea, if all the cases are in towns near the river, and the outbreaks spread down the river, this would give you a hint that there might be something lurking in the river water. Investigation might prove this suspicion wrong. It might be a disease-bearing mosquito that was being spread down the river, or something else. At least you could start testing some hypotheses.

There is a pub in Broadwick Street in London's Soho, just along from Carnaby Street. Called the John Snow, it does well serving ales to epidemiologists, anaesthetists (Snow gave chloroform to Queen Victoria during her last two confinements), statisticians and curious scientists. Just a little further along the street, there is a replica of a pump, shown earlier in this book.

The original pump provided local people with water when Broad Street was there. German bombs destroyed the area in World War II, so it is now Broadwick Street, and local shopkeepers hush you if you say there was once an epidemic of cholera there. One must not alarm the tourists!

Tourists need not be alarmed. The outbreak was a long while ago, 1853, to be exact, and in ten days, 500 people died of cholera in an area just 250 metres across. All the cases seemed to be in one small part of London, and they seemed to centre on Broad Street. Snow targeted that street—and he had previously suspected that cholera was spread in drinking water.

The Broad Street pump took water from a well, but some houses near the pump were free of cholera and some cases were located quite a long way from the pump. By careful questioning, Snow found that some people from further off preferred to get water from the Broad Street pump because it "tasted better" than other pumps. These were the *only* distant people to get cholera. Then he found that the nearby people who used the pump but were cholera-free were tea drinkers, who never actually drank unboiled water from the pump.

'Infection' meant something different then. An infectious disease was one caused by infecting poisons, and on that model, boiling could work by destroying the poisons. It would be some years yet before people knew about bacteria causing disease, but the match between pump and cholera was a convincing one, so he called for the removal of the pump handle, stopping people from using the water. Within days, the link was broken and the epidemic was a thing of the past.

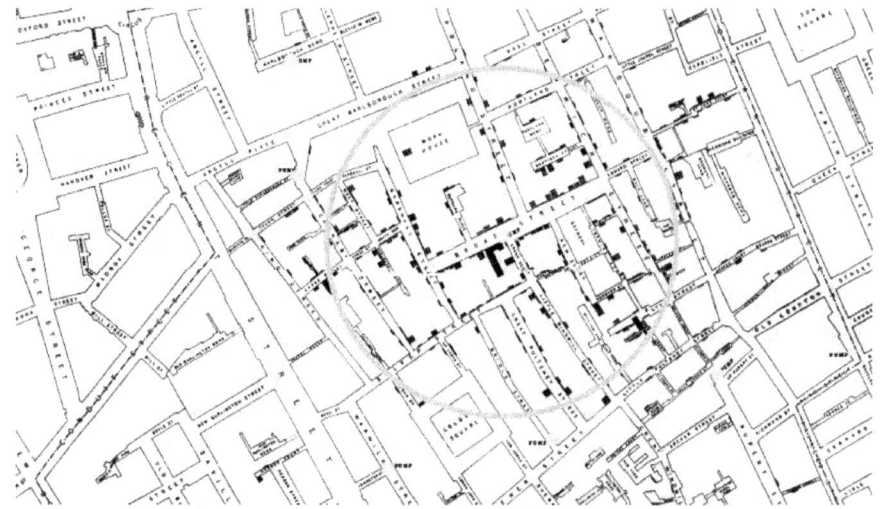

Snow's map of Soho: the available pumps are yellow dots, the blue circle includes almost all the cholera cases (black dots), and it is centred on the Broad Street pump.

We are told he drew and published a map showing the pattern of cholera cases near Broad Street. Each black dot was a house where somebody got cholera, and the pumps were shown as well, but the map was almost certainly drawn later.

The outbreak started on August 31, three days after a baby at 40 Broad Street developed cholera. This was the house nearest to the well, and the toilet for this house was a cess-pit, just like most of the other toilets in London. All human waste from the house went into this hole, right beside the well.

Cholera gives people the "runs" (or diarrhoea, to give it its proper name). Cholera germs breed in victims' stomachs, and are flushed out with the "runs". This puts them into the cess-pit, and then into the well beside the cess-pit. Groundwater moves in when water is pumped out, so leakage was inevitable.

Epidemiology needs subtle and creative minds and the clever use of science and even more so, of statistics. The only visible difference between malaria-carrying mosquitoes and some related mosquitoes which don't carry malaria is seen in the hairs on their legs. Only careful and systematic microscopy can reveal details like that. Graphs, spreadsheets, maps and tests can reveal more. Truth and honesty are essential or statistical analysis is a waste of time.

The World Health Organisation figures on the SARS outbreak show that in every other country with a major SARS outbreak, the mortality was 14 to 15%. The Chinese admitted to just 6.4%, and that stinks of manipulation. In Hong Kong, the mortality rate lifted at the end, as old cases died and no new cases arose, but in China, the death rate did not show this final upward flick.

Somebody in China was still lying. Without absolute honesty, epidemiologists stand little chance in any future outbreak of disease. SARS was stopped in 2003, but only just, and it may not happen this time—or next time.

Water wars and poison wells.

Water is profitable and the first half of the 21st century will see many fights and skirmishes, maybe even wars, all about water supplies and drinking water, but already we can see how water supplies that were public property are being forced into private ownership to satisfy the economic fetish of some 'Free Market' dogmatist living far from the consequences of that privatisation.

One of the more alarming horror tales of foreign aid, aimed at helping poor people comes from the push to sink many tubewells in Bangladesh. Before this time, most water came from contaminated surface water which commonly had high bacterial loads. Afterwards, many Bangladeshis were drinking water which well and truly exceeded the WHO guidelines for dissolved arsenic.

By the time the problem started to become apparent in the early 1990s, some people had been drinking arsenical water for as long as twenty years. This raises the question: why hadn't anybody noticed the arsenic? The short answer is that arsenic accumulates, and the victims become ill slowly, with none of the symptoms of classical poisoning that you might see with a massive dose of rat poison. There was slow, equally lethal damage to selected organs of the body, and a variety of cancers. In rural Bangladesh, the level of medical support was insufficient to detect the pattern.

In the developed world, deaths are looked at closely, and so are certain sorts of disease, and the results are placed on computer and examined closely by epidemiologists who are seeking patterns. That luxury is not available in the underdeveloped world, and the tell-tale patterns did not show up on any screens.

To make the issue harder, only tubewells drawing water from a depth of 20 to 60 metres tend to carry heavy loads of arsenic, and even then, the dosage can vary markedly—and as the water table rises and falls, levels change.

You can cover up a one-off mistake, but not an ongoing poisoning. 87% of Bangladeshis now have access to a tubewell within 150 metres of their homes, and there may be as many as 10 or 11 million tubewells. The problem had to be faced: a bad mistake had been made.

The mistake was a reasonable one, as the previous water sources were faecally contaminated surface pools, and about a quarter of a million children were dying each year from water-borne disease. That, the experts said, was killing too many. Groundwater was clean (of germs), and it had been used in the past, but it was from shallow 'dugwells' that took only recent rainwater as it sank down.

Deeper down, the arsenic has still to be flushed in the 20 to 60 metre range, though below 60 metres, the older sediments are flushed and largely safe. Against that, if there is extensive pumping from lower levels, arsenical water will sink down to replace what is pumped, and some cases have been reported where previously 'clean' wells are now showing higher levels of arsenic.

So where does the arsenic come from? Geologically, Bangladesh is made up of sediment carried down from the mountains to the north. Groundwater drifts slowly through it, making its way to the sea. Some of the sediment is old, some of it is recent: the older sediment has had most of the arsenic washed out, while any arsenic in surface sediments has long since interacted with air in times of drought, or organic matter, and turned into soluble arsenic that has leached away.

10: Rocks that were once alive.

The object in last shot on this page was never alive, but it's still a fossil.

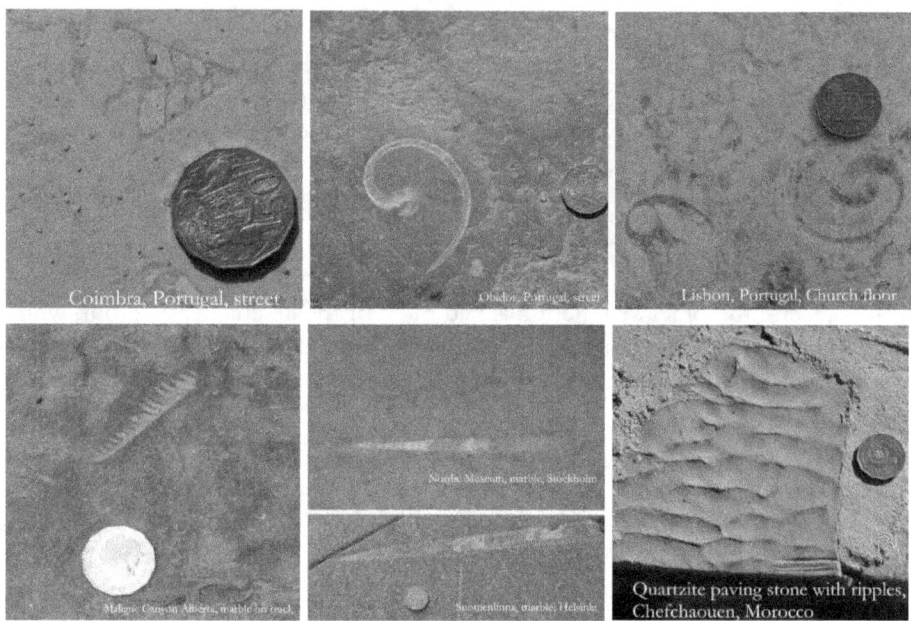

A quick primer on fossils.

Fossils are traces of old life forms which need to be interpreted, allowing for changes in death, and the warping caused by the compression of sediments. The best fossils are formed from living things when the material of a live organism is replaced by other material that is fine-grained and slow to deposit.

    There are many fossil types: some are formed when something rots away, leaving a mould that can be filled by minerals in groundwater, seeping slowly in. Chemical replacement happens when bones are buried, and over time, the mineral material of the bones is replaced by other chemicals, even opal, which is dissolved silica.

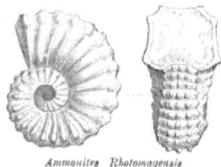

Ammonites Rhotomagensis

Frontispiece from *The Student's Elements of Geology*, 1871.

Decayed plants may leave phytoliths as traces, and these can be recovered from deposits and used by archaeologists as hints on past climates and crops.

Later, we will look at fake fossils. Here are some real fossils, to get us started.

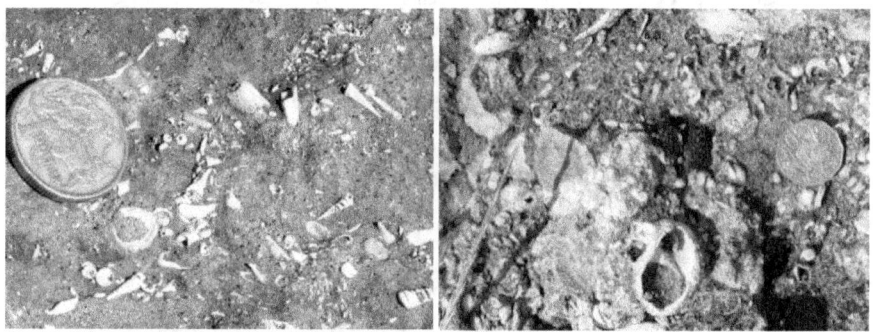

When shell grit and sand become stone, this is what we get. On the left, Fossil Bluff, northern Tasmania, on the right, Mandurah, Western Australia.

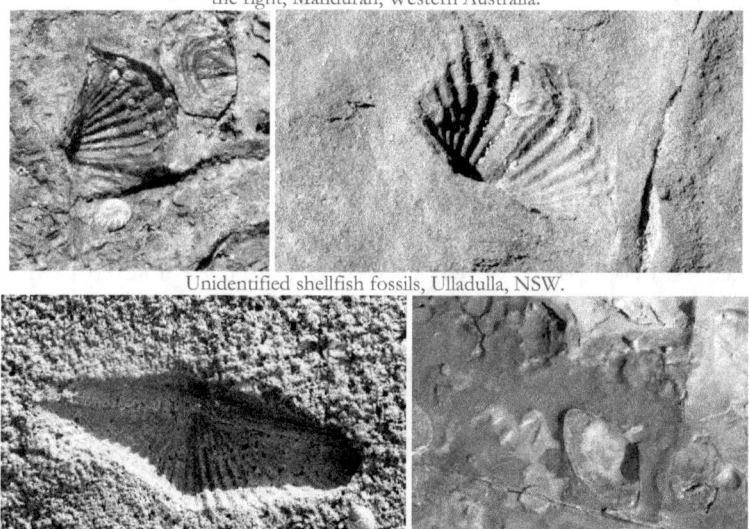

Unidentified shellfish fossils, Ulladulla, NSW.

Fossil lamp shell and unidentified bivalve, Ulladulla, NSW.

When sediments are laid down, we don't assume that the same bed will stretch over the whole of a depositional basin, but we do expect that where the bed occurs, it will have similar properties, so we should recognize the same bed if we come upon it somewhere else. And because we accept the Law of Superposition, we know (as well as a scientist can ever "know" anything) that the beds will always turn up in the same order.

By relating different beds to each other in different places, geologists can build up a very reliable "stratigraphic column", which shows how all of the beds would have appeared if they had all been formed in the same place.

This has two main uses: we can find the continuation of a particular bed that carries useful minerals in it, and we can build up a plan which shows the total sedimentary deposition that we might have got in one place, if that place had stood still and collected all of the sediments that have fallen all over the world since the Precambrian era. This was to have important implications for geologists in the 19th century as they tried to use the stratigraphic column method to get an age for the earth. We will come back to that method later.

When people ask me what my hobbies are, I say I have just one: lighting fires. No, not bushfires, though I used to light bushfires for research, under very tightly controlled conditions. (I was a research assistant with senior scientists, we had tankers, up to 30 crew, knapsack sprays and tools, and the work we did gave rise to the present six-point bushfire danger scale.)

The later editing of this book took place in New Zealand, and thereby lies a tale of lighting a fire in the hearts of three anonymous children. One February Saturday, my wife and I decided to take in the Auckland Art Gallery, a superbly designed building with delightful Jura Grey limestone floors and stairs from Bavaria, all highly polished.

Now as the people who travelled with me in 2019 in Spain, Portugal and Morocco can attest, I keep my head down when walking on marble or limestone, watching out for fossils like the ones seen here. When I photograph fossils, I usually lay down an Australian 50 cent coin for scale: these coins are 32 mm across. People often only see me picking up the coin after getting my shot, and if they question me, I have been known to claim that I am a one-trick (and not very profitable) wizard.

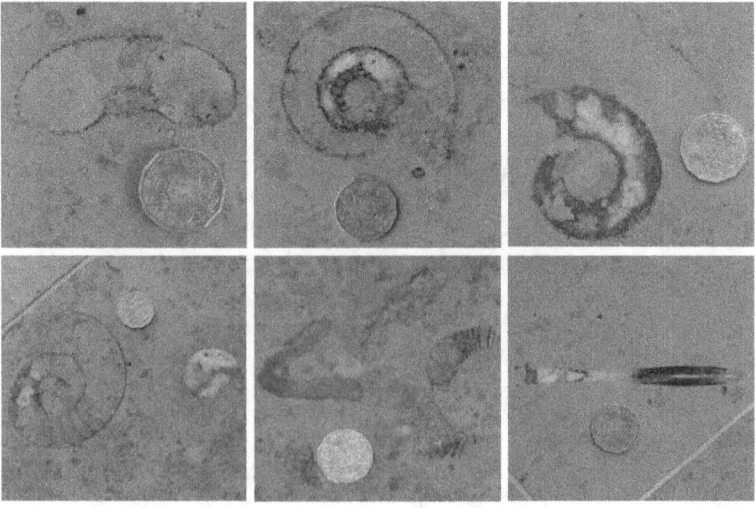

Five ammonites from the floors of the Auckland Art Gallery, and a single belemnite.

This is fun, but real joy is discovering that the floors and stairs of a building are all fossil-stuffed limestone. True joy is having a father and three children ask what I was photographing. Sheer blissful joy is walking out the front door of the Auckland Art Gallery, an hour later to hear the oldest one, a girl of 8 or 9, giggling with glee at finding yet another fossil in the outdoor paving.

I think I won that one. So yes, my hobby is lighting fires, and that's why fossils matter.

On April 29 1962, President John Fitzgerald Kennedy addressed a gathering of 49 Nobel laureates at the White House with these words:

> I think this is the most extraordinary collection of talent, of human knowledge, that has ever been gathered together at the White House, with the possible exception of when Thomas Jefferson dined alone.

It isn't the sort of comment we might reasonably expect to hear from (or about) the most recent incumbent of the Oval Office, but Jefferson (1743–1826) really was an amazing intellect. At a time when religious leaders mainly denied that fossils were traces of ancient life, and kept pushing the Biblical 'truth' of a 6000-year-old planet, Jefferson analysed the science, and proved that Noah's flood could never have happened.

The proof is in Jefferson's *Notes on Virginia*, which also reveals his lively interest in fossils, making him an early starter, but not the first to realise what fossils were. Around the same time, that eminent French scientist we know as Buffon, was on the same track, saying that "marbles, limestones, chalks, marls, clays, sand, and almost all terrestrial substances" came from oceanic sources.

Buffon wasn't the first, either: the beginnings of fossil awareness lie in the year 1666, when two fishermen caught a large shark off Livorno (Leghorn) in Italy. The Grand Duke of Tuscany, Ferdinand II, had the shark's head sent to Nicolaus Steno (1638–1686).

Steno dissected and drew it, publishing his findings the next year. It was a straightforward anatomical drawing, but if it was a Great White Shark (as some people think), then the drawing was rather less than accurate. In fairness to Steno, the fish was probably in poor condition by the time he received it.

Steno noted that the shark's teeth resembled unusual stones called *glossopetrae* which came from the island of Malta. Pliny the Elder had said these curious stones had fallen from the sky during lunar eclipses.

Others called these the tongues of serpents which had been turned to stone by Saint Paul when shipwrecked on Malta in CE 59. This legend was the origin of their name, which literally means *tongue stone*. It was a pretty legend, and there was a thriving trade in tongue stones, which were sold as antidotes for poison.

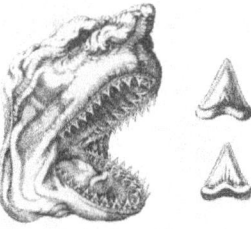

Steno's shark, and its teeth.

Dip one in a glass of poisoned wine, buyers were told, and the poison will be detoxified. Gullible rich buyers swallowed this and paid well for a tongue stone amulet that could be dunked in any suspect glass of wine (which they also swallowed), but Steno's scientific comments rather damaged that trade.

He said the tongue stones were shark teeth, which was a reasonable thing for an anatomist to say, but his next step was to wonder how the tongue stones got to where they were found, deep inside rocks, especially if the world was only a few thousand years old, as most people then believed.

We will look at the age of the Earth in chapter 14, but perhaps Steno saw he was on dangerous ground here, a Protestant in a Catholic country, soon after the religion-based Thirty Years War. The shark teeth had been enclosed in rock that showed an impression of the teeth. Clearly, he said, the rock had formed around the teeth, and they had not, somehow, been inserted into the rock.

He also concluded that if marine fossils like shark teeth are found high on mountains, those mountains must once have been under the sea. For some reason, perhaps safety, Steno converted to Catholicism in 1667 and two years later, he published what is usually called "his Prodromus". A prodromus is an introduction, a prologue, but he never went any further with it.

In 1675, Steno was ordained as a priest, and in 1677, he was appointed a bishop, and left geology behind. In England, Robert Hooke and botanist John Ray had also said fossils were the remnants of organisms, but as a bishop, Steno could make this challenging science more acceptable in a changing church.

Still, his major contribution came before he joined the church, when he spelled out his basic principles of geology which spread fast: by 1671, there was an English translation available. That translation was less than helpful, because it was entangled in old thinking which was valid at the time, but which is far from valid today, so here is a modern version that conveys the two laws and the two principles that Steno left for us. We met these earlier, here they are again, briefly.

- **The Law of Superposition**: any stratum is younger than the strata on which it rests, and it is older than the strata that rest upon it.

- **The Law of Original Horizontality**: strata are deposited horizontally and then deformed to various attitudes later.

- **The Principle of Lateral Continuity**: strata initially extend sideways in all directions.

- **The Principle of Cross-cutting Relationships**: anything that cuts across layers must post-date them.

Steno also noted that there are two major rock types in the Apennine Ranges near Florence. The lower rocks have no fossils, while the upper layer is rich in fossils. In keeping with the thinking of his day, he took the lower layer to be rock laid down before the creation of life, while explaining the fossils in the upper layer as relics of Noah's flood. He was wrong, of course, but for the first time, somebody had tried to point to geological evidence of different periods of the Earth's history.

There the matter stayed, with a few minor attempts until Baron Georges Léopold Chrétien Frédéric Dagobert Cuvier (1769–1832) made a major contribution to science by explaining how the whole of an animal might be reconstructed from just a few small fossilised parts:

> Every organism forms a whole ... if, for instance, the intestines of an animal are so organised as only to digest fresh meat, it follows that its jaws must be constructed to devour a prey, its claws to seize and tear it, its teeth to cut and divide it, the whole structure of its locomotory organs such as to pursue and catch it; its sensory organs to perceive it at a distance...

This more than made up for Cuvier's earlier mistake of interpreting extinct animals as clear evidence of a series of catastrophes. He fell into this trap because he accepted Buffon's estimate for the age of the Earth, no doubt augmented by the story of Noah's flood.

Cuvier specialised in the reconstruction of vertebrates from fossils, and he knew each layer of rock beneath Paris was identified by its fossil contents, wherever that rock outcropped on the surface. For the first time, Cuvier and his assistant, Alexandre Brongniart were able to draw up an idealised stratigraphic column where every stratum was assigned a fixed position. All of a sudden, people realised that fossils were important, and rushed to apply Cuvier and Brongniart's methods to the geology of other areas.

Cuvier wanted to change natural history from an anecdotal hobby into a rigorous science, just as firmly founded as chemistry and physics. To do this, he needed a range of specimens to study, and he collected all that he could for his museum. In 1812, Cuvier broke, once and for all, the idea that there was a 'chain of being', a line of animals from lowest to highest (with humans on the top of the heap). Instead, he saw four equal assemblages of animals: the *vertebrates*, the *molluscs*, the segmented invertebrates (worms, insects, crustaceans etc., called the *articulates*) and the radially symmetrical animals such as sea urchins, which he called the *radiates*.

These, said Cuvier, were different and equally successful body plans, with no suggestion that the vertebrates should go 'on top', as in the old idea of a chain of being. Cuvier's four groups are still the basis of the modern phyla, although the articulates and the radiates have been further subdivided since his time.

Others, however, tried to link these separate branches together, and hoped desperately to find 'missing links' to join the branches. The platypus was seen in this light, linking the separated mammalian and reptilian lines, but with none of the evolutionary notions we give to a 'missing link' today.

To explain some of the facts of animal and plant distribution, Cuvier suggested there must have been land bridges which later disappeared. Further, when species were split into the New and Old Worlds, they were separated, and degenerated in different ways, giving rise to modern 'species'.

Fossils come in many varieties, ranging from the body impressions of fish in a dried-out Devonian period billabong (we will come to that story later) to flat carbon prints of outlines, opalised skeletons, footprints and tracks, fossilised droppings (coprolites), and rarely, even the original material, like the bones in California's La Brea Tar pits, or the insects trapped in amber which were so important to the plot of the movie *Jurassic Park*.

Most fossils are chemical replacements, atom by atom, so the fossil bone you see in a museum is really a piece of stone, where each atom has been washed away, over long periods of time, with rock atoms left in their place.

 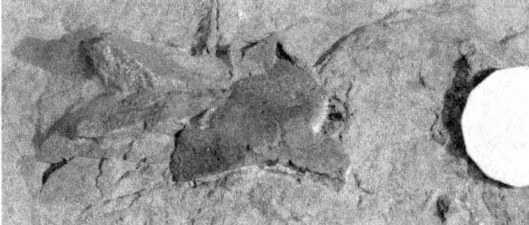

Glendonite trace and mould, Ulladulla, NSW. These are replacement fossils.

Glendonite is an interesting form of calcium carbonate that only forms below 5°C, in a cold ocean. So when we find forms like the specimens below on a rock platform, we know that the rock was laid down in a cold sea, because the hole on the left is where a glendonite crystal was later dissolved away.

In the right-hand shot, we see a piece of iron oxide that has formed when the crystal dissolved away, and later, iron oxide was carried in and laid down, filling the mould that was left behind. That explains some odd cobbles on the same rock platform, which is at Ulladulla, south of Sydney. There must have been a glacier somewhere nearby, calving icebergs that floated off, melting slowly and as they did, dropping big round pebbles that plunged down into the soft mud.

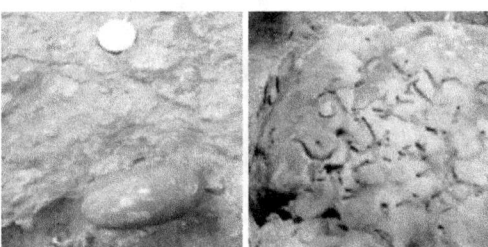

Dropstone, Ulladulla, NSW (left) and sandstone with "worm" burrows, Sydney (right).

In a sense, those dropstones are fossils of the past, but they are still the original atoms, near enough. Above, you can see a dropstone and also some sandstone that appears to have been thoroughly burrowed. One geologist friend used to think these holes were traces of fossil tree roots, but I tend to think they were burrowing animals of the wormy kind.

One other fossil type occurs as original material; the fossil placoderms of the Late Devonian reefs of the Gogo Formation in the Kimberleys of Western Australia. These fish fell off their reefs into deep quiet waters when they died, and were rapidly buried in a lime-rich mud which encased them and then set hard as limestone, holding the entire fish skeleton together, or the whole fish piece, for some were scavenged, either as they sank, or after they reached the sea floor.

The main point is that the actual bones are original, not a chemical copy of them, and more importantly, the bones were protected, just as a broken arm is, when it is put into a cast. Where most fossils are flattened and distorted, these were preserved, ready for removal. Preparation is easy: repeated washing in 10% acetic acid, air drying, followed by a plastic impregnation, more acid, and so on.

The secret of becoming a fossil is to die in the right place. Most dead bodies are eaten by scavengers, the bones are picked at by birds and small mammals, and bits are often dragged large distances. If you want to be a complete fossil, you need to avoid all that. You need to be buried in oxygen free conditions, in fine sediment, deep enough so you are protected from animals that may eat you.

This rarely happens, and this is why the story of human evolution is mainly built on partial skeletons that have been gnawed at by hungry animals, keen to get at the rich and nutritious brain tissue, with the main bones crunched to expose the marrow in the long bones, and some of the rest trampled by passing animals.

Finding fossils.

Collecting fossils became a popular obsession among the middle classes in the 19th century, exposing more people to evidence of ancient life. Mary Anning (1799–1847) was a self-trained professional by the time she died, far too young.

Living near Lyme Regis on the southern coast of England, she found and prepared fossils for rich collectors and museums. She is even commemorated in a verse that was once an assertion about the reality of fossils, though most of us only know it today as a tongue-twister:

> She sells seashells, by the seashore
> And the shells that she sells, are seashells I'm sure.

Anning's story has been picked over, romanticised and fictionalised, but there can be no doubt that she was respected. When she died of breast cancer, her obituary was published in the *Quarterly Journal of the Geological Society*, an organisation that would not admit women as members until 1904.

Anning's father worked as a carpenter and cabinet maker but he was also an enthusiastic fossil collector, and she made her living from fossils after he died in 1810, selling them to rich middle-class enthusiasts who were beginning to take an interest in at least the more genteel aspects of science.

In 1811, she found an ichthyosaur, now in the Natural History Museum in London. Later, she discovered the first plesiosaur, and in 1828, the first British pterodactyl, and rich people flocked to learn from her how she did it. One visitor, Lady Harriet Silvester, wrote in her diary that Anning had reached the peak of her profession at 25 "by reading and application".

> ...she has arrived to that greater degree of knowledge as to be in the habit of writing and talking with professors and other clever men on the subject, and they all acknowledge that she understands more of the science than anyone else in this kingdom.

To the educated middle classes of Britain, Anning needed no explanation:

> The violence of the weather lately washed down...and exposed a mass, which, on digging out, proved to be the vertebrae of some animal, whose size must have been enormous. It is in excellent preservation, every process and part being perfect...Many are the conjectures with respect to the animal; some imagine it to be the gigantic buffalo or the rhinoceros, and others the elephant. That intelligent osteologist, Miss Anning, of Lyme, surmises it to belong to either the behemoth or the hippopotamus, yet admits that it far exceeds their acknowledged dimensions.
> —*The Gentleman's Magazine*, December 1824, p. 548.

The cliffs of Lyme Regis were then, as they are now, unstable, and the art of the fossil collector involved walking along the base of the cliffs at low tide, seeking out fallen rocks while inspecting fresh exposures.

Her best finds are probably still not credited to her. She sold many fossils to collectors, who often later donated the specimens to museums, though the records show only the donor's name, not that of the person who did the hard work. She learned on the job, but without her efforts, scientists like Richard Owen would have lacked access to the fossils which showed them clearly that the Earth was far older than anybody had thought.

Owen graduated in medicine, but took a post at the museum of the College of Surgeons, and met Georges Cuvier in 1830. After studying in Paris, he became superintendent of natural history at the British Museum in 1856, where he was at first supervised by the principal librarian in an organisation that cared more about art works and books than about biological specimens.

Owen is generally considered to be Cuvier's successor, but he found certain advantages in being British. There was a long-running and major disagreement about the status of the Australian platypus, a fur-covered, milk-producing, egg-laying animal with an apparently reptilian skeleton. Owen cultivated the curators at the Australian Museum in Sydney, and remained the only European scholar with easy access to a good supply of platypus specimens.

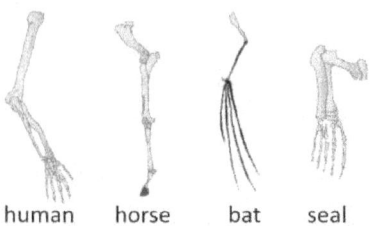

human　　horse　　bat　　seal

The principle of homology: these are all pentadactyl (five-fingered) limbs.

Owen was the first to recognise and use the idea of homology. A bat's wing and a seal's flipper are called homologous because they have the same underlying pattern. (We need to distinguish this from analogy: the wing of an insect is analogous to the wing of a bird, but these have different origins, and so they are not homologous.)

From the 1820s, dinosaur fossils were found in England, and by 1841, Owen had used the term 'Dinosauria' to describe these animals. Sadly, he behaved in a scurrilous and underhanded way to try to undermine the explanation of evolution put forward by Charles Darwin. He was not a nice person.

At the same time, Owen's recognition of a moa from a small fragment of bone, like his statement that *Archaeopteryx* was "unequivocally a bird" were just two of his contributions which consolidated the Darwinian model of evolution. In time, everybody was fossil-aware, and even novelist Charles Dickens joined in:

> … implacable November weather. As much mud in the streets as if the waters had but newly retired from the face of the earth, and it would not be wonderful to meet a Megalosaurus, forty feet long or so, waddling like an elephantine lizard up Holborn Hill.
> —Charles Dickens, *Bleak House*, London, 1852, 1.

The Cambrian explosion.

The standard story is that life on earth emerged about 3.5 bya, ±10%. Then after some 3 billion years of no real progress, about 542 million years ago, most of the phyla of modern animal groups appeared quite suddenly in the fossil record.

Rocks forming before that time were classed as 'Pre-Cambrian', and the rocks from 542 to 488 million years ago are from the Cambrian period. The massive outburst of life at that time is what we call the Cambrian explosion. During that time, life forms changed fast, probably due to increased competition.

One plausible explanation for the 'explosion' is that the evolution of vision forced animals to develop better ways of avoiding or resisting predators and catching prey. Certainly, during and after the Cambrian, all sorts of multi-celled life forms left traces of their hard body parts behind in the rocks.

Later, came the discoveries of exquisite fossils of soft-bodied forms in the Burgess shales of Canada, and the Ediacara rocks of South Australia. Clearly, what happened at the start of the Cambrian was the evolution of animals with hard parts that were more easily turned into fossils.

So it was a boom, rather than an explosion. In the Vendian period that came before the Cambrian, (~650 to 542 mya), the soft-bodied animals multiplied in number and form, ready for the start of the Cambrian, but what triggered the second half of the boom, the 'explosion'?

Was it an asteroid? Humans always want catastrophes like a rock that plunged into the Earth's atmosphere and changed the world forever, but we can see the seeds of change in the 20 million years of soft-bodied fauna that flourished during the Vendian, before the changes.

The evidence favours a slow build-up of changes in the Vendian, leading to a locally-driven catastrophic (but disaster-free) change which produced the hard-bodied beasts of the early Cambrian. There is no need for any *deus ex machina* cataclysm, riding in on an asteroid, the Four Horsemen of the Apocalypse, all on board a hurtling rock. There are almost as many theories as there are palaeontologists, and one (or more) of those notions may well be right.

Preparing a fossil.

Sometimes the rock can be cleaned away with dilute acid, after a long soaking, but at other times, you need to work with a needle and brush, clearing away the surrounding rock. It took Raymond Dart several years of delicate work with a sharpened knitting needle before he was able to free the jaw of the Taung skull from the cranium, but that was extreme. (We come to the Taung skull shortly.)

A typical specimen will usually take several hundred hours of preparation. Dart, by the way found that skull in a box of rocks sent to him from a limestone quarry—it hurts to think what else from that quarry went to make cement.

Museums around the world share specimens, and often make casts of their best examples to sell to other museums, so what you see will often be a copy. It will have been cast from a mould of the original, and will be painted to resemble the original. The best casts are made by painting layers of latex onto the prepared surface to get a mould that can be peeled off.

Dr Alex Ritchie, preparing a latex cast of one of the fish slabs at Canowindra, Australia (see next).

Prices vary, and museums often share or swap material, but good specimens of Australopithecine skulls typically cost about US$400, part of which goes to the museum responsible, to help fund future expeditions.

Devonian billabong: a case study.

First came the billabong, then the fish, and last of all, the people. Other nations call billabongs 'oxbow lakes', but Australians call them billabongs. Nobody uses oxbows any more, and a billabong is an unlake-like, silent, murky backwater, abandoned by its river, deserted, left to its own devices.

The forces that shaped Australia in the past are the forces shaping the land today. Devonian period Australian billabongs formed then, just as they do now, all over the world, wherever rivers wander across flat plains. You can even see billabongs in Siberia when you fly over them in spring, their dark shapes marked by late snowdrifts along their banks, you can see them in north America, and, of course, in Australia. Only the name varies.

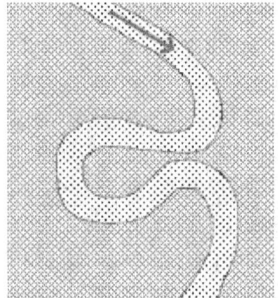 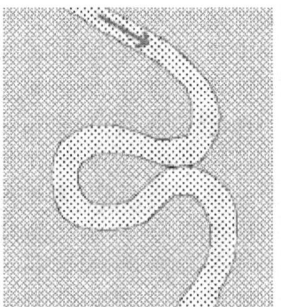 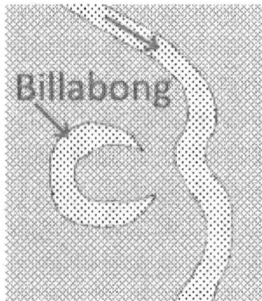

How a billabong forms. The river flows from top to bottom.

A billabong starts when a river twists in ever-larger meanders. One day, two nearby loops come so close that rising waters break through, finding the river a shorter path. Later, when the waters fall away, the new opening is fixed, and the cut-off winding arc of river bed remains as a long sluggish pond.

After a billabong forms, the river banks block both ends, and the billabong is cut off. It is still below the local water table, so even in dry weather, it is full of seeping water and it fills to the brim when flood waters pour over the plain.

Close to the river, billabongs carry many of the same species of plant and animal. The billabong resembles the river, but develops its own special ecology, disrupted each time the river floods, reduced in each drought when the billabong shrinks, and re-established each time the rains return.

The Devonian climate was probably much like today's climate. Droughts in Devonian Australia must have happened as they do today: only the plants and the animals were different, 360 million years ago. Since then, there have been 360 once-in-a-million-years droughts. In those searing droughts, the water remnants feel dry to the touch, and the rivers become chains of warm mud puddles.

Devonian droughts would have seemed even worse than today's, for most of the animals were fish, creatures of the water. There were no reptiles, no birds, and no mammals to walk or fly away, for these creatures had not yet evolved, and the fish were trapped.

Devonian waters teemed with many kinds of fish, but mainly with placoderms, strange armoured fish who protected their front halves with plates of bony armour, while leaving their tails exposed. On the land, a few insects had followed the first adventurous plants ashore. Then there were the lobefish, the crossopterygians which not only breathed through gills but also gulped air into a sort of primitive lung.

In the warm oxygen-starved mud, those primitive lungs could be a real advantage. So could the lobed fins that these fish may have used to drag themselves slowly through the mud, heave and flop, heave and flop, after insects. Maybe they could also escape to some other body of water, but in a drought, that would be a slim chance.

These were the animals of hope for our human future, because their descendants would eventually give us the amphibians, and through them, the reptiles, the birds and the mammals of today. Those lobed fins tell it all, for they have bone structures just like our own arms and legs.

But that lay well ahead, and was no future for the animals of Canowindra's Devonian drought, where nothing escapes and all life is forced, gasping, closer and closer to the centre of a shrinking tepid mudbath, pushing and shoving for a share of the dwindling, soothing cover of water.

Inevitably and suddenly, the end came for the stick-in-the-muds. The last muddy water drained away or evaporated, and the remaining fish, pushed into a smaller and smaller area, choked to death by the thousands. Finally uncovered, their skins scorched and burnt by the sun as they flapped their last, they were trapped, unable to heave and flop their way to the river nearby.

Soon after, the rains came. A small, soothing flood trickled over the ground and into their billabong, covering their bodies in a gentle shroud of sand. The weight of the covering sediment pressed the fishy remains down into the mud below. More floods followed, sea levels changed, continents rose and fell.

In time, the billabong was covered with perhaps two kilometres of sediment which hardened into rock. Time passed, modern fish, reptiles, birds and mammals evolved far above, and slowly, the oppressive burden of the rocks was eroded away from the fish, still lying deep within the rock.

In the last two hundred thousand years, the fish have been close enough to the surface of the land for the effects of weather and groundwater to reach them. Slowly during that time, in the spaces where the fishes' bones had been buried, the enveloping rock turned into a clay-like paste.

More time passed, more rock was worn away. Humans came to the land. In 1956, a ratepayer complained about a dangerous bump on a bend in a local road. His shire council sent out a bulldozer to rip several large slabs of rock from the bend to ease the way. Some of the fish were free of their burden at last.

The bulldozer driver was a shrewd observer. He saw strange patterns on one of the slabs, and carefully turned it over, meaning to do something about it later. Time passed, rain washed away some of the clay. A local beekeeper saw the slab, and what looked like reptiles in the rock. He wrote to the Australian Museum in Sydney, saying the slab was to be found near Canowindra, a small rural town, about 300 km from Sydney.

It was some time before a palaeontologist could visit the area. When he got there, he was so excited, he wanted to camp beside the slab to protect it from theft. Eventually, he was persuaded that nobody would steal six hundred kilograms of rock that had already lain there for many months, but he slept restlessly in town that night.

The slab was carefully cleaned and prepared, and by 1966, it was on display at the Australian Museum in Sydney. Then in 1993, a curious alliance of a local dentist, a museum palaeontologist, and sympathetic local councillors, arranged for a 20 tonne excavator, complete with driver, to see what else the site held. They had a dream.

The result was a huge array of slabs, carrying the prints of several thousand Devonian armoured fish. Each slab, prised up and turned over, bears the impressions of the tops of a swarm of fish, while the rock that remains behind carries the impressions of their undersides.

The lower surface was covered until it was ready to be unveiled later as part of their dream, a new 'Age of Fishes' museum. The top slabs were carried ceremoniously in proud council trucks to a shed at the Canowindra showground, to be cleaned by eager volunteers, and I was one of them.

Cleaning a fossil slab takes time and care. You start with a tiny chisel, hammering on the slab, picking at the embedded remains of a fish. Soon, the rock begins to crumble at the edges, and you move on. You brush away the fragments with a dry brush, scrub at the growing hole with a wet brush, and mop it dry. For finer work, you attack the slab with a toothbrush or a large needle, the sort once used to sew up wheat bags.

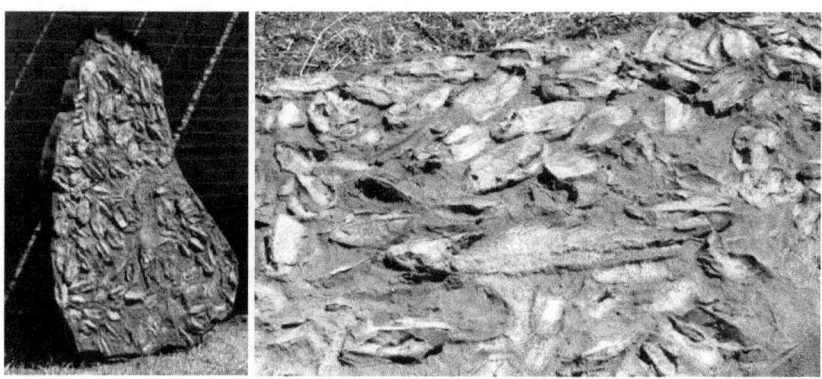

Two views of one of the Canowindra slabs.

All the while, you scan urgently for signs of the 'ornamentation', the pattern on the outside surface of the bony armour of the placoderm fish. Beside you, a set of drawings reminds you of the sorts of shapes you can expect to meet. Each blow, each gouge is a puzzle, a tiny gamble that you've read the picture correctly.

One way or another, the work was done, the fossils escaped from their rock, and went from the dusty showground to the Museum of the Age of Fishes in Canowindra, celebrating the Devonian billabong, and that tragic day, 360 million years ago.

Interpreting a fossil.

Some fossils are tougher than the rock around them, and they stick out of the rock as it weathers, while others are less tough, and weather away, leaving a small hollow in the rock. Fossil hunters need to look for these signs, but then they need to dig into the rock around the area, in the hope of uncovering some more finds, which then need to be cut out, whole if possible, and taken away to be worked on.

Thin carbon film impressions of fish and leaves often lie between a pair of shale beds, and weaken the rock so that, as you chop away at the rock, cutting down through the beds, a block will crack neatly at that point, one half of the fossil on each side of the new split.

In some cases, when an area is known to be full of fossils, explosives, drills and earthmoving equipment may be used to break the rock up into pieces that can be searched through more easily. While a few fossils may be damaged this way, the overall benefit in terms of fossils found before they weather makes the practice worthwhile.

Sometimes, scientists can do a remarkably good job of reconstructing a murder victim's face from a skull, and the same methods can also be used to reconstruct a fossil skull. Of course, they can never know what colour the animal was, but they can often make a reasonable sort of guess, based on similar animals that are alive today.

Interpretation is never easy, because many fossils are badly squeezed out of shape when the rocks are compressed, so museum preparators need to study taphonomy, the science of body decay. This gives them useful facts about the way animals bloat, and lie with their feet in the air. Dead wombats always lie with an ear to the ground until they bloat and again, later, they roll back over.

I am, let it be said, always willing to do a little more science. Hiking through a damp area on Kalianna Ridge, two of my children and I found some parts of a wombat skeleton. As well-brought-up children, they understood the need to experience taphonomy, so they dropped their packs and we spread out over the open but leech-infested ridge, gathering in the bones that predators had scattered.

Scattering is the usual fate of large animal bodies, unless they are quickly covered over. The bones which are not trampled before they are fossilised are quite likely to be trampled when they come back to the surface.

Bones of a dead wombat, retrieved from a thorough search over a 50-metre radius.

Today, we might argue that *Australopithecus* ("southern ape") was not the best name to give to an upright walking individual, even if it had a small brain, but when Professor Raymond Dart invented the name, he was trying not to draw too much fire upon himself.

The problem he faced was that his find (known as the "Taung child", from where it was found and its obvious youthfulness) was small-brained and far too many British scientists were convinced that the small-brained thing was wrong as an ancestor. Piltdown Man was our ancestor they said: he had a big brain, and best of all, he was found in Britain!

So they wasted no time attacking Dart, anyhow, but Dart was right, and they were backing a complete fake that some people already knew to be a fake, but let's begin at the beginning. In the film *2001: A Space Odyssey*, the memorable scene over the opening titles shows an ape discovering that things could be hit with a bone to great effect.

It was an idea that came originally from Dart, who argued that humans emerged from hominids who learned to use tools to hunt. But why did Dart think they used bones? He was teaching in South Africa in the 1920s. He was a medical man, but fascinated by fossils, and in 1923, he was sent two boxes of rocks that might (or might not) contain fossils. They arrived on a day when he was supposed to be getting ready to act as best man at a friend's wedding.

Peeking into one of the boxes, he found a brain-shaped piece of rock, a cast formed on the inside of a skull, lying loose. He knew right then that he held something important. This was an endocast, a copy of part of the inside of the skull, but it wasn't just any old brain—this one was special, because of its small size and its brain stem.

Keep in mind here that the people who interpret fossils work like Sherlock Holmes at his best. To those who can read, a glimpse of a document can be enough, but those who can read fossils can gain just as much from a single glimpse of just the right hint.

The cast fitted a block of stone in the case, so a major part of the brain owner's skull was there as well, encased in rock. It was only a partial skull, but he had the endocast, and that had already told him a delicious tale. Reluctantly, he did his duty at the friend's wedding, then rushed back to his rocks; *what had he seen?* In simple terms, he saw the stem of the brain, and that was enough to say that its owner walked on two legs, like us. Here is how he worked it out:

> I was also convinced from the earliest period of my investigations that these creatures had placed great reliance on their feet for walking and running and that, consequently, their hands must have been freed for other tasks. This was implicit in the globular form of the skull which was obviously balanced on a more vertically placed type of backbone

than that of a gorilla or chimpanzee. The improvement in the poise of the head implied a better posture of the whole body framework, since there must have been a relative forward displacement of the *foramen magnum* (the hole in the base of the skull which links the brain with the spinal cord).
—Raymond Dart, *Adventures with the Missing Link*, 1959, 11.

Brains fit neatly into the skull, so the endocast showed where the brain stem left the skull, at the bottom, like our brain stems, not at the back, like the brain stems of a gorilla or a chimpanzee. But if this was a primitive human, it must use tools, and where were the tools? Dart proposed that early humans lived in an *osteodontokeratic culture*. This hypothetical culture used tools of bone (the *osteo* part), tooth (the *donto* portion) and horn (which gives us *keratic*).

In the end, Dart's successors found plenty of stone tools at Olduvai Gorge, but let's look at Piltdown for a moment. Over a period of time, people digging in a gravel pit at Piltdown (about 20 km from Brighton in southern England) found a cranium (skull) and mandible (jawbone), a canine tooth and a sort of tool, an odd piece of bone that looked rather like a cricket bat. Remember that: it's a clue.

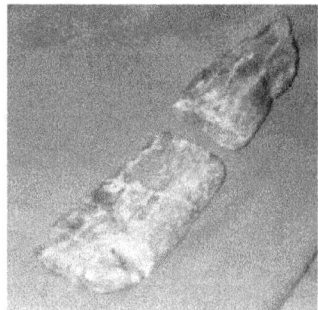

The Piltdown "cricket bat".

From the bones, they constructed a whole man. That is not unusual: fossil specimens are nearly always incomplete, but this one was a fake, though realising this took some time. The Piltdown Man may have been a hoax that went wrong, or it may have been a real fraud.

The first pieces of the puzzle were found in 1912, and the whole assemblage was not exposed as a fake until forty years later. During the time that scientists believed in "Piltdown", a lot of damage was done to science, especially in Britain, where the reconstruction was considered totally authentic. A number of American and French scientists were less convinced about its status.

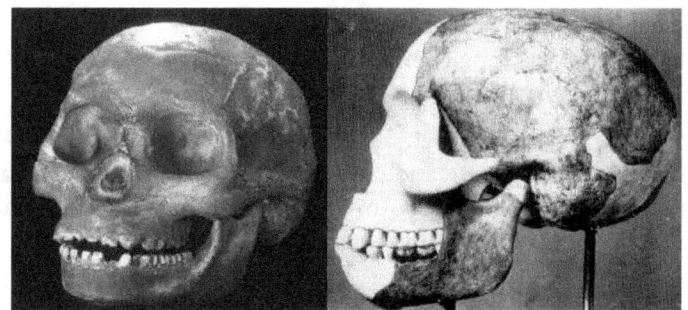
The Piltdown skull, reconstructed.

Piltdown man was not even a fossil. It was made up of some pieces of a modern human skull, and a piece of the jaw of an orang-utan. The bones had been broken, the teeth of the lower jaw had been filed, and all the pieces had been stained with chemicals. It wasn't even a *good* fake!

So how did it happen? After they were found, the pieces were taken, cleaned up, and then "reconstructed" into a whole skull. The doctored find provided just enough information to allow a reconstruction, where the jaw was completed to fit the skull, and the skull was completed to fit the jaw.

The faking was clearly done by an expert, who had carefully broken off or filed down any of the bits that would have given the game away, although the filing of the teeth was quite amateurish, and gave a surface which, in hindsight, is a dead give-away. It was almost certainly something the faker intended to be seen.

Somebody seems to have wanted this hoax to be found out. (I should note here that I handled the Piltdown material in 1993, and some of my comments are filtered versions of what I have learned over the years from various experts, speaking off the record. I brought to those discussions, and this discussion, some considerable experience both as a hoaxer and as an investigator of frauds.)

For one, Pierre Teilhard de Chardin knew more than he ever admitted. It was Teilhard who "found" another tooth for the jaw at Piltdown. We now know that this tooth was also a fake, and that it was faked in a different way (it was painted rather than stained with chemicals), almost as though somebody was trying to warn the original fakers that they were not the only fakers around.

That the tooth was painted and not stained like the skull and mandible was a fairly sure indication that it was faked by a different person. The giveaway is the "Piltdown cricket bat". This was crudely hacked out of an elephant's leg bone, made using a metal chisel, and then painted like the tooth and planted in the same deposits as the original find. It was as though somebody was saying "All right, you have your First Englishman: now here's his jolly cricket bat!".

The "bat" was a very poor copy of a cricket bat. It was too small, the camber on the back was wrong, and the handle was wrong. Many people did not recognise it as a cricket bat, and did not realise that it had been planted as a deliberate fake to make them think. Instead, they wondered out loud why there were no other bone tools like it found anywhere else.

Even when another Frenchman, a good friend of Teilhard de Chardin pointed out that the 'bat' had been worked with metal tools (not available until long after the supposed time of Piltdown man), people said nothing.

It looks to me as though Teilhard, a Frenchman who would have known little about cricket, "planted" the tooth to get a reaction from the fakers, then he (or somebody else acting with him) "planted" a poor copy of a cricket bat, and later got his friend to draw attention to the chisel marks on something supposed to have been made long before metals were discovered, but then gave up on the whole thing. The English were too thick.

The true status of the Taung child lay hidden inside its jaw until 1987. In both humans and the other apes, the "adult" teeth come into place in a specific sequence. There is one order of appearance in humans, and a different order of tooth eruption in the other apes.

Concealed inside the Taung child's skull, teeth were erupting, and if we knew how they were developing, we would know what the Taung child was, human or ape. As there is only one Taung child, you cannot slice it up, just to see what is inside. You can take X-rays, but there is too much other material in the way, and the things we are looking for are much too faint.

It seemed for many years as though we would never know what was inside the jaw. Then in 1987, two scientists named Glenn Conroy and Michael Vannier had a bright idea. Instead of cutting the skull into thin slices, they could make a series of virtual slices with X-rays, and feed all of the results into a computer.

The computer would then use a method called back projection to build up a three-dimensional picture of what was inside the bones, and show us how the Taung baby's teeth were erupting. In short, a CAT scan would give the answer.

So the researchers took their series of X-ray shots, just 2 mm apart, in three different dimensions: vertically, from front to back, vertically, from side to side, and horizontally. (They called it the sagittal, coronal and transaxial planes, if you prefer the technicalities.) The detail is less important, though, than what they found, because the answer was delightful: "the Taung 'child' is not a little human, but just as important, it is not a little ape."

The Taung baby is a betwixt-and-between, a half-and-half, a missing link if you wish, and we would never have known if the two researchers had not decided to give it a CAT scan! Sadly, we had to wait more than sixty years to find out what it was.

The disappearing petrified trees.

Australia has lots of "lakes" or "lagoons" along the coast of New South Wales. These are and were highly productive areas for humans. An early missionary minister Lancelot Threlkeld stayed with the Awakabal people on a reservation on the large Lake Macquarie, south of Newcastle in the 1820s and 1830s.

While he was there, they told him about "Kurra-kurran", the local name for a petrified forest sitting in the waters of the lake, and he wrote about the area in 1834. An early Australian geologist, William Branwell Clarke, visited the area in 1842 and described the petrified trees there. He also mentioned others at Reid's Mistake, where water from the sea flows into and out of Lake Macquarie.

Threlkeld reported the local legend behind the "stones": they were part of a rock hurled down by a giant and angry goanna (monitor lizard). Clarke told them the scientific culture's counter-legend, which was published in the proceedings of the Geological Society of London for 1843, and an abbreviated version appeared in the local press in 1845.

Clarke estimated that there were 500 petrified trunks in the lake. By the 1970s, there were only 30 left, the rest having been "souvenired", and there are now just a few still visible. According to geological gossips, many of the previously collected specimens were dumped in the lake by somebody from the Lake Macquarie Council.

The scientific version of their origin is that these trees were killed and buried by the hot ash from a volcanic eruption, probably from a volcano erupting somewhere well off the present eastern coastline.

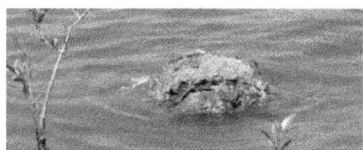
Pushed-over petrified tree trunk, Lake Macquarie, NSW, Australia.

Recently, I found just two, which is not a Good Thing. The Blackall Park one, however, was hard, and the area is so pillaged that I won't say how to get there:

if you have a special need, contact me with an email address for a reply with how-to-get-there details. The persistent will find me readily enough through social media.

Clarke noted that many of the stumps appeared like those of recent trees. The tops showed clean horizontal sections, suggesting that they had not been broken off. He had already decided that these trees were pine-like, and there was no way that modern pines could be sheared off that way. This is a fair assessment: The stumps weren't alone, he said, pointing to the examples at nearby Reid's Mistake, as well as a number of more distant examples.

To be precise, the survivors are at Swansea Heads (on the southern side of Reid's Mistake), where the trunks are embedded in stone and largely immune to theft or vandalism, either official or unofficial.

To get there, wait for low tide and then drive towards Swansea, take a left towards Caves Beach, but pick up Northcote Avenue, then Lambton Parade and park at the end in the Reid's Reserve car park. Find a track through the small dune to a beach on the east (it's very sheltered by a reef, so often has lots of children and dogs). Go to the water's edge, swing right and head around, out onto the rock platform.

 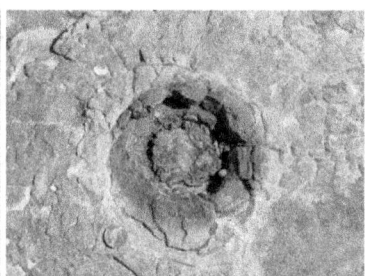

Petrified trees, Swansea Heads.

I found Clarke's description of the area, which he calls by its local name of Kurrur-kurran, in the *Sydney Morning Herald*, 4 August 1845, p. 2.

Fake fossils from Morocco.

Morocco is, if you will pardon the geographically inaccurate cliché, a Mecca for fossils, and you will often see them exposed for sale like the left-hand shot below, where the cat is walking mainly on old ammonites. The sawn-open ammonites that youths will offer to sell to you are too hard to fake, and too easy to find. Many of the others are fakes.

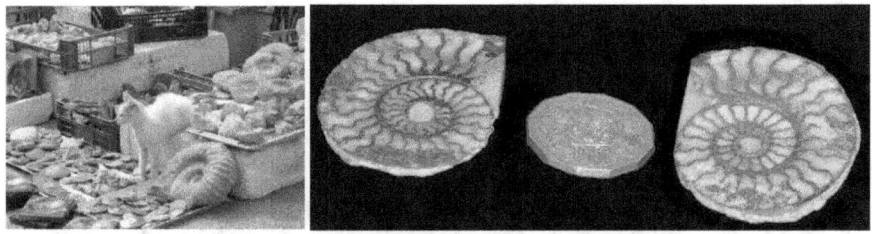

Genuine fossils, on sale on the left in Casablanca, purchased on the right.

The problem is that many of the fakes are offered for sale outside Morocco, and it is possible that the shops where I have seen them in the USA and Australia really knew no better, but all it takes is a careful look, and a tap of a finger-nail. Stone goes 'ching!' but resin makes a dull thud, by comparison. So when you try tapping on a moulded resin "fossil", the feel and sound is quite different.

A fake fossil 'swarm', made in a factory in Morocco, with real fossils glued onto a stone base.

The main fakes are trilobites, and the copies are getting better and better. Experts I have talked to describe it as an "arms race" between the detectors of spurious fossils and the fakers, who hang out on the internet forums where the secrets of detection are revealed. One good rule: if it looks too good, it's probably a fake. The next two pictures, below, show things that are most definitely fakes, though they are made with real fossils. Here, the show is just too good to be true.

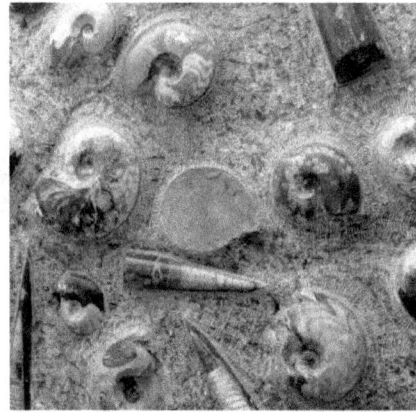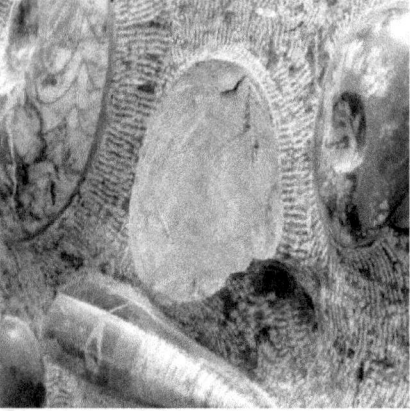

How genuine fossils are attached to a stone base, using a thick layer of epoxy glue.

The fossils are glued to a heavy stone slab. Look for the thick layer of epoxy glue that attaches them (if they don't fall off). Watch out also for alert shopkeepers who hate chortling photographers.

Trilobites are the most common fakes, and you will almost certainly end up with a resin cast unless you keep your wits about you. Start by buying a cheap ammonite pair, sawn in half, from a teenager on a street in a small town.

These are too cheap to be worth faking, but they are hugely over-priced in shops. Then spend some time, tapping your genuine ammonite with a finger nail until you know the feel and sound.

Wayfaring with fossils.

Fossils were once anything that was found buried, including archaeological material, and this confusion lives on in some quarters, as this true tale will reveal.

> In 1993, I was working for the Australian Museum as an educator, but I had been seconded to an aid project in the Pacific (training science teachers), and I brought home a pickled Giant African Land Snail.
>
> Knowing these snails ought to be a dubious import into Oz, I declared it, outlined the preservation methods used, explained that it was destined to reside in the Australian Museum, and I was told "You know more about this than I do. Take it."
>
> Clearly, I had pre-empted the argument phase, but equally clearly, I act honestly. That had to change in late April that same year, when I was forced to get an export safely out of Britain.
>
> A colleague at the museum, knowing I would be in Edinburgh, had asked me to courier home a type specimen of a Devonian fish that he had lodged with the Royal Museums of Scotland in Edinburgh some years earlier.

I agreed, took the right letters to effect the loan, collected the fossil in bubble wrap and cardboard and set off, with the appropriate documents saying what it was, and that it was a proper and legitimate loan.

All was well until I arrived at Heathrow on my way out. As I had all the papers, and worried that there might be questions if I did not show what I had, I declared the fish. The customs officer spoke impeccable English but let us just say that he was clearly of non-British parentage and culture. "You can't take that," he said.

I asked why not. "You don't have the appropriate EU documentation, and you are stealing our heritage." Now I might have said, "Mate, this is a Scottish fish, I'm a genetic Scot, and it's being loaned by a Scottish museum to a Scot in Australia—and he found it in the first place. So just whose heritage are we talking about?"

I didn't, because that is not my way. I could see the surreal aspect, but I was diplomatic. It availed me nothing. He still demanded that I surrender the object as yet uninspected, which lay in my carry-on bag. I argued, and explained that it was one of a kind. It was a type specimen, and special to science. He produced an alphabetical list of sciences on a sheet of paper. "Is this archaeology or zoology?"

"Neither," I said. "It comes under palaeontology."

He bristled, sensing that I was being smart or worse, and scrutinised his sheet of paper. Because I'm good at reading upside-down, I pointed to the word he was seeking. "It means 'study of fossils'," I said.

I had, however, recognised that this gentleman, for all that he was polite, well-informed on paperwork and probably well-meaning, was either untrained or as thick as two bricks—or both. I concluded that this was not somebody who could be trusted to look after a type specimen and treat it with due care.

A type specimen is very special to scientists: it is the original specimen from which a name was given. My colleague had found this fossil, described it and named it, but now it needed some further study, which is why he was borrowing it.

I had, in the same bag, a piece of partly metamorphosed shale from Wales, wrapped in newspaper, so I brought that to the top of the bag. It was, after all, of no real value to me. Let me emphasise that I did not, at any time, imply that this was the fossil because I am, after all, a totally honest man. *I spoke no lie.*

His hand darted out, seized the package, and tipped the stone out of the protective wrapping so it landed with a bang onto the counter (confirming my assessment of him), turned it over, dug at it with a grubby thumbnail (further confirmation), then handed it back. "I suppose you can take it," he told me, somewhat reluctantly.

I thanked him nicely, wrapped the Welsh shale again, placed it protectively over the fossil fish in its bubble wrap, and walked off.

I said nothing until I was home, when I told an edited version of this story in a British weekly, *New Scientist*, as part of a discussion on rare fossils being sent through the mail, the risks they were put under, and the need to have customs officers examine them. The officers needed to be educated first, I said.

My colleague examined the fossil fish as necessary, and it was later couriered safely back to Edinburgh, though not by me. I was lying low, and having outed myself, I have avoided Heathrow ever since, because they may still be waiting for me.

A practical scientist.

Georges Cuvier argued that the whole of an organism has a pattern that is defined by its way of life in a predictable way. We met him earlier, saying that the different parts of an organism tell the same story, and this brings us to an anecdote of an exchange that I suspect never happened—but it should have!

In this yarn, Cuvier was visited one night by a joker, dressed as the devil. "Cuvier", roared the prankster as he burst in, "I've come to eat you up!"

The man was reading in bed, and he looked up calmly to consider the figure. "You have horns", he declared, "and a tail, and a cloven hoof, so you're herbivorous, and you can't!" With that, he returned his attentions to his book.

11. Making money from rocks.

Worn stone and sawn stone, Fairy Bower, Sydney

When people seek riches from the planet, all bets are off, especially when it comes to getting rich without hard work. The people who used 'hydraulicking' in California, Australia and New Zealand probably thought they were clever, but the Romans did similar things in Spain, almost two millennia back.

Malakoff Diggins, California (left) and Las Medulas, Spain (right). Ruins from gold mining.

Any good history of the rise of technology must deal with several phases. To get past a stone-based culture, technology needs a place to work, and that generally means a village with places to store stuff. Then there is the need for materials, most of them drawn from the ground.

The next stage is finding methods: smelting, writing, agriculture, measurement, engineering, but to build a civilisation based on technology, humans need concepts like force, energy and interdependence, all tied together by the art of communication. Mainly, though, our ancestors needed the right materials.

There were seven metals known in ancient times: gold, mercury, tin, lead, silver, copper, and iron. Iron, on rare occasions, found as the metal in meteorites, but mostly, people who wanted to get some metal had to treat an ore in some way, to smelt it, to get the metal.

So the future technologists needed to know where to dig, and that was the start of economic geology, but sometimes what they needed was a bit of cunning. A Pole named Seweryn Korzelinski reached the Victorian goldfields in the 1850s, where he noticed four Irishmen living in two tattered tents. He saw that these men spent two days a week digging and the rest of the time drinking and fighting. A three-pint bottle of brandy cost £1, so he guessed they were getting gold. He sank a shaft close to their claim and it paid off.

Later on, shafts went far deeper, but the first arrivals in a gold rush took the easy gold that lay in the gravel in the creeks. After that, there was more gold hidden in the soft clay beneath the gravel. This sort of mining was called "surfacing", but as the surface gold ran out, the diggers had to change their methods.

They dug down into the top layers of a light but variably coloured soft shale called the pipeclay, but what made them start going down past the pipeclay and digging really deep holes? Most of the gold seekers were practical people, operating on a limited budget, and once their money ran out, they would have to sell up and go home. They were unlikely to risk their money in sinking a shaft, just on chance. So what sort of hopes, what knowledge did they have?

During the 1850s, many of the shafts went 60 metres straight down. These were driven down through sediment—sand, soil, small rocks and clay, not through solid rock, though later on, shafts would also be cut through rock.

Soil was easier to dig than rock, but it was still hard work. When miners dug through sediment, their shafts needed "timbering" or "slabbing". That meant cutting and adding timber frames to stop the sides collapsing, and there was also 200 tons of earth, or more, to be removed from the hole.

If gold turned up at the bottom, everybody was happy at their good fortune, but if the hole missed running into gold, the diggers had wasted a small fortune in sinking and reinforcing the long shaft.

After a while, people noticed that some patches were richer than others. No rule could ever be established as to the occasional large nuggets, but then somebody must have noticed that where gravel came in contact with the shale, there were pockets that were full of gold.

In the 1850s, there were people who called themselves geologists, and ordinary folk could buy books on 'geology', but nobody had too much of an idea about what lay beneath the scenery. Geology was still an infant science. Fossils were rare wonders, volcanoes were poorly understood and earthquakes were hard to explain. There was no real idea of such a thing as geological history and that made the rocks hard to understand.

Coal miners would dig out flat beds of coal sandwiched between layers of sedimentary rock. They would follow the coal seam until a geological fault ended it, or the bed thinned out. Then they looked around for a logical place to continue, but the geology of gold was much more complicated, and a good vein might be anywhere. It was a mystery, and it was a secret.

Ordinary diggers had no idea that far below the ground they walked around on, there were old land surfaces and old creek beds, ancient country, buried beneath deep layers of sediment.

When the shaft-sinkers began to find gold-bearing "drift beds"; they learned to trace "leads" through the drift bed until they struck a "gutter"—an old creek bed. By that good luck, a pattern was set. Looking back with modern knowledge, we can see that the first lucky diggers must have dug blindly into an old and buried creek bed, still with the load of gold it held when it was first covered over.

Victoria's 19th century rush yielded 2500 tonnes of gold, but that was less than 2% of all the gold ever mined in human history. Geologists say this may be less than half of what is still there in Victoria. Another 5000 tonnes of gold may lie in old valleys and gullies, buried far below beneath the flat and featureless plains.

Preparations were made in the late 1990s to combine seismology and drill cores to see what is down there, but no useful results seem to have flowed from this. Many of the gold deposits could be hundreds of metres down, and out of reach but as a science writer who was once a botanist, I knew there was another way of tackling the issue.

Some of the shallower deposits might be found with a bit of cunning, using the science of *geobotany*. This idea is by no means new, and the best reference is more than 50 years old, though the practice is much older even than that.

The first use of plants as indicators of economic ore deposits happened in 1828, the method was revived in 1837 in Germany, and immediately after World War II in the USSR, as it then was. One plant, the blue-flowered *Ocimum homblei* only grows in soil containing 100 parts per million of copper or more.

Observing the distribution of *O. homblei* has led in the past to the discovery of several ore deposits. The only plant growing in high-gold areas known so far is the horse-tail of British Columbia. Still, prospecting with plants may be worth studying—though not worth investing in, at this stage.

There was a better and related solution hinted at in a 1986 editorial in *Science*, when Philip Abelson discussed the way plants extract gold with their roots. This gold can be detected by neutron activation analysis, when levels as low as 1 part per billion can be found in 10 grams of wood ash.

This approach was being used in Canada in areas where most of the solid rock is covered by a layer of glacial till. Some trees have roots that reach down to the bedrock and take up gold, if it is present. The ash produced by incinerating tree samples taken over gold deposits gives readings of around 100 ppb, while trees away from a deposit will have about one third of that reading.

The snag has always been that sampling in the cold Canadian winter is challenging, and in summer the flies eat you. The answer is aerial sampling. These days, with GPS, it would be easy to hover over the trees, taking samples with a grab dangling from a helicopter or a drone, tagging each sample with an exact location that could later be used to generate a map of places where gold levels were high. This would show where to drill cores.

Now think for a moment of another place where potential gold deposits are covered by a layer of sediment. What if a similar plan could be applied to the Murray valley? Our daughter, like her parents, is trained in botany, and she mentioned that Australia's *Eucalyptus* trees can be quite efficient at accumulating gold from the soil. So economic geology can still be expected to head off in some odd directions.

Using the rocks.

Who built the seven towers of Thebes?
The books are filled with the names of kings.
Was it kings who hauled the craggy blocks of stone?
And Babylon, so many times destroyed.

> Who built the city up each time? In which of Lima's houses,
> That city glittering with gold, lived those who built it?
> In the evening when the Chinese wall was finished
> Where did the masons go?
> —Berthold Brecht, 'A Worker Reads History', *Selected Poems*, 1947.

The first stone tools were made during or soon after the brief lifetime of the Taung child. Raymond Dart may have been right about people using bones, teeth and horns first (or not: it remains an interesting suspicion). Still, a couple of million years back, give or take a bit, somebody noticed that certain rocks broke to make sharp edges when they were banged together.

Before long, somebody found that the shiny rocks like flint and obsidian worked best. Somebody else saw that glancing blows gave longer, thinner, sharper flakes, and technology began. Science, the art of explaining events and predicting effects, may have taken a little longer, but the first hints were there, way back then.

Modern humans (us) are inventive and inquisitive. For more than 30 millennia, they have shared ideas and techniques. Those new notions spread from valley to valley, as people visited their neighbours, or when traders came and carried new methods away. War parties might take prisoners, captives who may have ingratiated themselves (or saved their lives) by demonstrating a new skill. Ideas spread, one way or another.

Whatever tools they used, communication was one of the earliest things humans did. Early communication may not have involved speech as we understand it, but the tool-makers of Olduvai Gorge, *Homo habilis*, must have taught—or shown—each other how to make tools, two million years ago.

Many people working on human evolution think it was the tool-making which made *Homo habilis* into the first humans. Yet working stone is an art, and if the making of tools was to be more than a flash in the pan, the art of tool making had to be passed on, from parent to child. Communication must have gone hand in hand with tool-making.

About the same time in many places, humans started to produce art. Cave paintings at Lascaux and Altamira, rock art in Australia, carvings in Czechoslovakia, decorated pottery, and more, and it started a new trend.

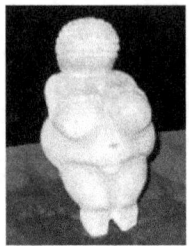

The 'Venus of Willendorf' or 'Willendorf Woman' is held in the Naturhistorisches Museum Wien, or Natural History Museum in Vienna. It is carved from oolitic limestone, and is typical of other 'Venus' figures.

In Australia, people may have been keeping records on rocks even earlier, using both rock engravings and painting on rock surfaces. These should be thought of as teaching aids, intended to illustrate parts of the rich cultures that held communities together. The rock art was more library than art gallery. Myths, legends and teaching stories of many sorts are held in those grooves on flat sandstone.

As the left-hand illustration below shows, the grooves actually started out as pecked holes that were joined up by rubbing with hard stone. The white invaders in 1788 brought deadly diseases, some violence, and a great deal of cultural disruption: that engraving was probably started before or soon after 1788.

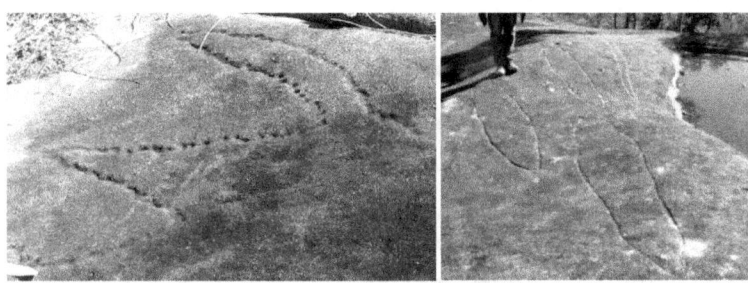

Australian Indigenous rock engravings near Sydney.

Notice how, in the right-hand shot above, some foolish person has scratched over the outline. This should *never* happen, any more than people should walk on the engravings. (That happens as well.) Paintings in an art gallery should never be touched. The same applies to art engraved or painted on stone surfaces!

Australian Indigenous rock engravings near Sydney.

There are two safe and good ways to make engravings show up well. The first is to go out in winter or early or late in the day, when the sun is low in the sky. The other is to carry extra water and pour it on the engraving. If you don't have a scale like the one on the right-hand shot, get somebody's foot in (not on) the picture.

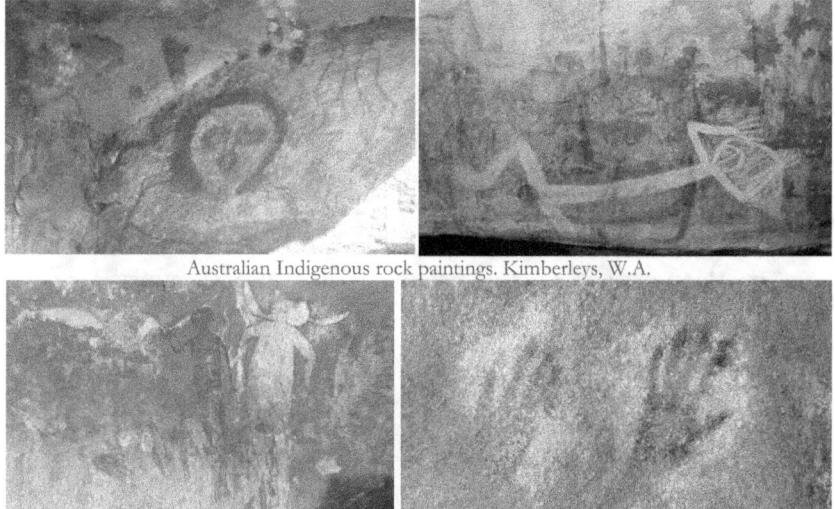

Australian Indigenous rock paintings. Kimberleys, W.A.

Australian Indigenous rock paintings, Mitchell Plateau and Malka's Cave, W. A.

Much of Australia's prehistory relates to Indigenous art painted with mineral pigments in rock overhangs. There are also engravings on rock surfaces in some places, and occasional stone tools that can tell a story. The paintings are information-laden, but how do you date them? After all, mineral pigments offer little material for carbon dating.

From fossil records, we know that Australia's megafauna died out, about 45,000 years ago, and the 'Tasmanian tiger' or thylacine lived on the Australian mainland until about 4000 years ago, when the dingo was introduced, and probably out-competed the thylacine. A painting showing megafauna is unlikely, but there are certainly sites where thylacines can be seen in paintings in northern Australia.

> In 2001, Richard Roberts of the University of Melbourne reported on his attempts to date 28 'young' megafauna sites (to sample only the later part of the lifetime of the species) where megafauna remains had been found across Australia. As the sites were older than the 40,000 year limit for radiocarbon dating, he needed to rely instead on uranium-thorium and optical methods to date the sites.
>
> The uranium-thorium method is used to date calcite bands in caves that lie above and below animal remains, while the optical dating method relies on the effects of stray radiation on quartz grains. These radiation effects promote electrons in the quartz to higher levels, but while the quartz is on the surface and exposed to sunlight, the electrons are continually released. Once the quartz grain is buried, the energized electrons can accumulate and provide a 'clock' indicating how long the grain has been buried.
>
> Using these measures, the sites returned dates in the range 51,200 years down to 39,800 years, but while there is still no 'smoking gun', people definitely appear to have been involved. The effect may, however, have been indirect, with Indigenous people using fire as a tool to clear and manage the landscape, and in the process, depriving the megafauna of the cover and food they needed.

We know that about 30,000 or 40,000 years ago, humans started to produce art in Europe, and by 31,000 years ago, they were painting the Vallon-Pont-d'Arc caves in France, and producing beads, figurines, and more. The human remains that we find with this art show us the makers were not Neandertals, but the really astounding thing is the range of the art.

The Cro-Magnon people often lived in 'caves'—meaning under rock overhangs on the sides of valleys, places probably better described as rock shelters. Many of these shelters were well above the valley floor, and would have needed some sort of a ladder to get people up to them.

The Cro-Magnons got their name from one such rock shelter near Les Eyzies in the Dordogne of France, where five skeletons were found in the Cro-Magnon shelter in 1868 during railroad construction work.

We know the Cro-Magnon people used the rock shelters, because they left signs of their occupation behind them. They left rubbish, bone scraps, bones with engraved sketches on them, remnants and scraps left over from the making of stone tools. One shelter at Les Eyzies, Abri Pataud, was completely "dug", and yielded more than 50,000 worked pieces of flint.

These were our people, the ancestors of the scientists who would come later, but in other parts of the world, people used rock in other ways.

The Egyptians used stone to write hieroglyphics, to make statues like the Sphinx and tombs (The Great Pyramid at Giza). [Christine Macinnis].

Like cultures all over the world, they used stone to make homes for the dead, long before they made homes for the living. In Asia, stone was carved into art:

Stone carvings, Angkor Wat.

Stone carvings, Angkor Wat.

A Roman bridge in Spain (left), and (right) a convict-built bridge in Tasmania.

Temple of Concordia, Agrigento, Sicily, 5th century BCE, and a nearby wall.

Deserted toilet, Kanyaka, South Australia (19th century), Kamares aqueduct near Larnaka, Cyprus.

The aqueduct was an early means of transporting water with no energy cost, but it required complex masonry and the ability to make waterproof channels. We think of aqueducts as a Roman invention, but there were earlier ones in Assyria, and they probably derive from the *qanats* of Persia (chapter 9). The Aqua Appia, the first Roman aqueduct, was built in 312 BCE.

A surviving (but not operating) aqueduct, Segovia, Spain.

The aqueduct would work by gravity if builders could survey a suitable route, build supports and construct a waterproof channel to carry water without leaks. While the most spectacular part of an aqueduct consists of the arched bridges over valleys, these were massive challenges, and avoided whenever possible. For the most part, the aqueduct was a water-proof channel that was laid in a ditch, winding around the hills.

Whether the channel was a spectacular bridge or a simple ditch: the channel had a smooth concrete base and side walls of stone slabs or masonry, covered with a layer of *maltha* cement, a mixture of lime, pork fat, and the rubbery milk of unripe figs. This surface was totally impervious to water.

Finally, the top was covered over with vaulting, a semicircular stone arch. The standard form was a channel 1.2 metres wide, and the height of the channel was 1.8 metres. If you ever visit a surviving relic like Pont du Gard, look for the tunnel which disappears into the hill on the Nîmes side of the aqueduct.

Pont du Gard was a huge structure, but it had to be built because the valley it crossed was filled with a raging torrent each spring. When this sort of expense was not demanded by regular floods in the valley, the Romans took their aqueduct down there in a lead pipe called a siphon, which was able to carry the water under pressure, and deliver it to the same height on the other side of the valley. This proved a great deal cheaper than bridging the valley.

Rocks as notepads.

This book started in a discussion on Sydney's North Head, and because the area is my playground, it is appropriate to look more carefully there. In the 19th and 20th centuries, the area housed Sydney's quarantine station, and some of the healthy people held there, carved messages in the sandstone.

Engravings on Sydney sandstone, North Head, Sydney, Australia.

How geology can shape a battle.

The Battle of Chickamauga was the second bloodiest battle of the US Civil War between the South (Confederate) and North (Union) forces. In a sense, it was the South's 'last hurrah' but it also spelled the end of Union General William Rosencrans' military career.

The battle took place on rugged terrain along Chickamauga Creek in northwest Georgia near Chattanooga, Tennessee, and the geology of the battlefield and surrounding areas played an important role in the complexity and outcome of the battle.

Geologist Stephen W. Henderson explored the geology of the Battle of Chickamauga some years back, and explained that the Appalachian Mountains in the Chickamauga area are made up of limestone, sandstone, and shale, and the key is that these rocks weather and erode in different ways.

Some rocks formed valleys, while others formed ridges. To engage in a military campaign, generals had to manoeuvre troops around these ridges while looking for passes or gaps to get through.

Henderson said that when the Union army moved south after they captured Chattanooga, they were chasing the Confederate forces under General Braxton Bragg into a place in Georgia called McLemore's Cove which is a plunging anticline. This is formed by a rather non-resistant combination of dolostone and limestone which opens to the north and is closed to the south, where the resistant sandstones of Lookout and Pigeon Mountains come together.

Bragg tried to trap the Union army in the cove, using the geology, but that didn't work. The Union troops, which had been split into three wings, were able to come together near Chickamauga, to the north of McLemore's Cove where the valley is broader. The Confederates tried unsuccessfully to drive the Union troops back down into McLemore's Cove where it would have been easier to defeat them.

Within the battlefield there are six different Ordovician rock formations, various dolostones and limestones. It turns out that the higher ground was high because of variations in the rock type.

The topographic crest where the Union Army had stationed themselves was limestone which had chert in it and it was the chert that caused it to be higher. So the Union line occupied a rather subtle high ground. Militarily, the high ground made a difference because the Confederates would have to attack up hill. The battle raged for two days.

The second day of battle was September 20, 1863. The Union troops were outnumbered but they held the strategic position on the high ground. Then when Rosencrans mistakenly moved some of his men out of their battle line, he created a gap that the Confederates were able to break through.

Part of the Union Army retreated to Chattanooga. General George Thomas rallied the remaining Union forces and made a stand on Snodgrass Hill, composed of a resistant dolostone with chert in it.

He managed to save the army from 25 separate Confederate attacks and then successfully escaped to Chattanooga. For his stand on Snodgrass Hill, Thomas was forever afterwards called the Rock of Chickamauga.

12: Thinking about rocks.

There were no men on board the fleet who had any knowledge of useful sciences, such as botany, geology, mineralogy, and natural history; and consequently there was no means of ascertaining the resources of the country, and applying the knowledge to the wants of the settlement.
— G. B. Barton, *History of New South Wales from the Records*, Charles Potter, Government Printer, 1889, 36.

... a geologist should be well-versed in chemistry, natural philosophy, mineralogy, zoology, comparative anatomy, botany; in short, in every science relating to organic and inorganic nature.
— Sir Charles Lyell (1797 – 1875), quoted in *A Thousand and One Gems of English Prose*, selected by Charles Mackay, 1872.

In short, everything that is now known concerning the configuration of the floors of the oceans proves conclusively that Wegener's hypothesis of continental drift is wholly untenable.
— Walter H. Bucher, 'The Crust of the Earth', *Scientific American Reader* (1953), page 67.

As compared with their present-day representatives, the Tertiary vertebrates were characterised by their larger size; not that small species did not exist, but that many which then lived were larger than any existing today.
— C. A. Sussmilch, *An Introduction to the Geology of New South Wales*, Angus and Robertson, 1922, 205.

To explain the observed phenomena, we may dispense with sudden, violent and general catastrophes, and regard the ancient and present fluctuations ... as belonging to one continuous and uniform series of events.
— Sir Charles Lyell (1797 – 1875), *Principles of Geology*, 1840.

Speak to the earth and it shall teach thee.
—*Old Testament, Book of Job*, 12:8.

Speak to a geologist, and it shall confuse thee—unless thou knowest how they think.
— Peter Macinnis.

Geologists spend most of their time trying to work out what lies underneath the surface of the Earth, or what history lies behind the formations which are seen on the surface.

They look for things which are happening today, and try to relate these effects to what must have happened in the past. They look for trends like the direction taken by a vein of minerals, and then try to work out where the vein will turn up again. This sort of reasoning is particularly important when a mine is located in faulted rocks, where a vein may appear to end abruptly, but continue on, on the other side of the fault, a few metres up or down from where it ends. More importantly, a clear understanding of how and when a geological deposit formed will tell us where to look for other similar deposits. An understanding of the traces which may be detected from the Earth's surface will save miners a great deal of unnecessary digging.

And most importantly, with 'Big ones', major earthquakes, due any year now in two of the world's seven largest economies, California and Japan, geologists are very keen to predict when earthquakes may happen, so they can reduce the economic disruption and the loss of life that will follow.

Geology even determines the economies of whole nations. The soils of Australia are old and poor, because the continental plate is too tough to buckle, so it won't raise mountains or start volcanoes, and the land is dry because there are no serious mountains to trap passing clouds and wring moisture from them.

So while Australia has some seasonal weather patterns, it has very little truly seasonal weather, and has much of the continent's weather controlled more by El Niño in the Pacific Ocean and the Indian Ocean Dipole to Australia's west. With a different geology, these climatic cycles might have a much smaller effect on the island continent of Australia.

The main assumptions of geology are Steno's law of superposition, with younger rocks lying on older rocks, together with the expectation that an intruding rock is older than the rock it intrudes into, that faults are younger than the rocks they divide, that included rock pieces are older than the rocks they are found in, and that more modern fossils are found in more recent beds, together with the belief that rocks containing similar fossils must have been laid down at about the same time.

Secondary assumptions include the view that there is a rock cycle, with material being raised to the surface by volcanoes, being weathered, eroded, washed and deposited in sediments which may later become sedimentary rock, or heated and compressed to metamorphic rock, or even melted to form igneous rock, which may later return to the surface again.

Until subduction zones were observed in plate tectonics, where one plate slides beneath another, the burial of rocks in the rock cycle was something of a mystery, but this is now much clearer.

Sydney's Triassic Hawkesbury sandstone turns up well to the south at Drawing Room Rocks.

Defining the shape of the Earth.

> 'Belief in Antipodes' became another stock charge against heretics prepared for burning. Some few compromising spirits tried to accept a spherical earth for geographic reasons, while still denying the existence of Antipodean inhabitants for theological reasons. But their numbers did not multiply.
> —Daniel Boorstin, *The Discoverers*, Dent, 1984, p. 108.

The first thing that had to be done before humans could go out into the world was to estimate the extent of the Earth. That meant, among other things, working out its shape. Most people say that before 1492, everybody 'knew' that the world was flat. They agree, it took the genius of Columbus to work out and prove that the world was round. Sadly, that's all complete rubbish, because Columbus was, to our certain knowledge, 2000 years too late! The ancients had seen, long ago, that the Earth cast a round shadow on the Moon. More importantly, there were other clear indicators that the world was a globe.

Parmenides of Elea, who may have been born as early as 540 BCE was probably the first to say that the Earth was a sphere, and he might also have been the first to realise that the Moon appears bright because it reflects the Sun's light to us, but there may have been others before him.

Around 520 BCE, a Greek named Anaximander suggested that the planet's surface is curved (he thought it was a cylinder), but by about 500 BCE, the followers of Pythagoras (an Italian Greek), concluded that the Earth was a sphere, and this was generally accepted.

In about 470 BCE, an Athenian named Anaxagoras (c. 500–428 BCE) said the Sun, Moon and stars were all made of the same material as Earth, and that the Sun is just a hot glowing rock. He was charged with "impiety", and this is easier to understand once you know that the Greeks believed the Moon and the Sun were gods.

The surviving old books seem to say that he either escaped or was banished—the story was a bit garbled. For good measure, Anaxagoras went on to explain, fairly accurately, why the Moon shines. He also explained lunar and solar eclipses, adding that the Sun is further away than the Moon.

So the ancients knew a thing or two, which makes it a bit odd that the Greek historian Herodotus, who lived at much the same time, doubted a story that he still shared with his readers, concerning some Phoenicians who had sailed all the way around "Libya", by which he meant what we now call Africa.

> These men made a statement which I myself do not believe, though others may, to the effect that as they sailed on a westerly course round the southern end of Libya, they had the sun on their right—to the northward of them. This is how Libya was first discovered to be surrounded by sea…
> —Herodotus, *The Histories*, 4.42

The Phoenicians were quite capable of making such a voyage, and clearly, the position of the Sun is just what you see on the southern side of the Equator, as Herodotus ought to have known. Anyhow, before Herodotus died, Philolaus had written in 455 BCE that the Earth rotates. Now we jump forward to Plato (c. 427 – 347 BCE), one of the more famous and influential philosophers.

As soon as you assume that the Earth sits still in the centre, and that the Sun, the Moon, the stars and the planets all orbit around it, you have a problem. As you watch the movement of the planets against the background of the stars, the planets seem to move forward and backwards.

Before the planets' movement was solved, some remarkably wild explanations were put forward. But before the curious wandering of the planets could be explained, people had to realise that, even if the planets seemed to wander about through the stars, there really *was* a pattern in their movements.

Leaving a lot of muddle out of the story, in 360 BCE Heracleides argued that Venus and Mercury orbit the Sun. He probably also believed that the Earth rotated on its axis. Eat yer heart out, Columbus! Heracleides actually said that if the Earth rotated and the stars stood still, this would look the same as the Earth standing still while the stars rotated around us. In the end, nobody took much notice, not even Aristarchus of Samos, who set the Sun neatly in the centre. Any real development had to wait for Copernicus.

We can blame Aristotle (384–322 BCE) for the delay. He blocked all progress when he proved by logic that the Earth did not move, and we need to examine this "proof". Everything falls towards the centre of the Earth, he said. If you move around the curvature of the planet, things still fall straight down, proving that everything falls towards the centre of the Earth.

Therefore, the centre of the Earth was also the centre of the universe, so therefore, the Earth did not move. End of discussion. And to a large extent, that was it, for most people.

In Aristotle's day, people could use an event such as an eclipse of the Moon to measure the curvature of the Earth. All they had to do was estimate the position of the Moon (or the Sun) in the local sky at some agreed point in the eclipse (totality for a solar eclipse, or first/last contact for a lunar eclipse) and then compare the position of the Sun or Moon with the result obtained at some (east-west) distant place.

That left one other avenue of study open: measuring the various distances and sizes. In 260 BCE, Aristarchus of Samos calculated the ratio of the Earth-Sun distance to Earth-Moon distance from the angles in an Earth-Sun-Moon triangle at half-moon.

It worked like this: the Earth-Moon-Sun angle at this special time was a right angle, and the angle from the Earth could be measured. With that, (and a bit of drawing in the days when there were no sine tables), there was your ratio. Getting actual estimates of distance would have to wait until Hipparchus came along, but at least it was a start—and we will return to this later.

Then Aristarchus moved on to calculating the distance and size of Moon from Earth's shadow during lunar eclipse, and his writings suggest that he was working on a Sun-in-the-middle or *heliocentric* model. Aristarchus' measures were way off: his calculations placed the Sun just 8 million km away (against a modern figure of 150 million km). Well, you can't win them all, but Eratosthenes (c. 276–c. 196 BCE) was about to get lucky.

Picture the world in about 200 BCE. In those days, "the world" pretty much meant the Mediterranean Sea and its surrounding land. The unknown empires of the Incas and the Aztecs, even the known empires in China, India and Africa did not count as "world" to Europeans.

Still, even with their limited view of "the world", every educated European believed the planet itself was round—a sphere in fact. The evidence all pointed to it, and besides, the sphere was the most perfect shape imaginable. So the Earth just had to be a sphere.

That meant there was none of this nonsense about falling off the edge of the world that ancient people are supposed to have carried on with, but the people of "the world", the Mediterranean one, had real problems when it came to measuring how big the planet was. They could not get a large enough tape measure, and even if they could, trees and mountains kept getting in the way!

Eratosthenes was another Greek astronomer, born in what is now Libya, and he died at Alexandria in Egypt. Among other things, he invented a way of generating prime numbers, which we still call the "sieve of Eratosthenes", and he was the first to propose using leap years to keep the calendar from drifting away from the seasons.

Because he was in charge of the huge library in Alexandria, Eratosthenes was well-placed to gather information, and he heard about a vertical well at Syenê (today's Aswân). There, on a certain day of the year, the Sun shone vertically down the well at noon.

On the same day of the year, the noon Sun was seven degrees and twelve minutes off the vertical at Alexandria. That was one fiftieth of a circle, putting the two places on the same great circle, but a fiftieth of the way around the globe.

Long before Eratosthenes, the ancient Egyptians had noticed this difference in the Sun angle. But they believed the Earth was flat, so they used the difference to calculate the distance of the Sun from the Earth. They said it was about 8000 kilometres away—as often happens, the result you get can depend quite a lot on the assumptions you use.

To somebody with Eratosthenes' education, the cause of the angle was obvious. He believed the Sun was a *long* way off, so the Sun's rays must all be parallel. That meant the angular difference between Syenê and Alexandria had to be caused by the curved surface of the Earth.

Now all he had to do was measure the distance from Syenê to Alexandria, multiply by 50, and there would be the circumference of the Earth. Unfortunately for us, Eratosthenes used the *stadion* as his unit of length. A stadion was the length of a race course, and the word is still in use today, when we talk about a stadium.

Back in the days when units were not standardised, the stadion was fine as a local measure for the people he was talking to, but every city had its own stadion, and they were all different. Sadly, we have no idea exactly how long *his* stadion was, but if we take the most probable length of the stadion, he was within a few percent of the correct size for the planet.

Then Eratosthenes said: "If the extent of the Atlantic Ocean were not an obstacle, we might easily pass by sea from Iberia [Spain] to India, keeping in the same parallel." If he had used Eratosthenes as his guide, Christopher Columbus would have known how far it was westward to the Indies from Spain. But the value which was generally accepted for the Earth's circumference in the 1490s was around 29,000 kilometres, when the true value is close to 40,000 kilometres.

With only a rough (and too large) estimate of how far it was, going *east* to China, and working on a world globe which was too small, Columbus expected that when he sailed west, he would find China about where America is.

Geophysics.

Once upon a time, geology was all about looking at the rocks and asking "Why?", but as the industrial age ballooned, geologists needed to ask "Where?". And so we got geophysics, which deals with the physical properties of Earth materials and the physical processes that determine the structure of the Earth as a whole.

The topic can include seismology, geomagnetism, and meteorology, as well as the study of large-scale processes of heat and mass transfer in the Earth and variations in the Earth's gravitational field, so this is a quick dance over the top: the main use of geophysics is to explore for valuable things.

Geophysical exploration methods include using variations in magnetic field, gravitational attraction, radioactivity levels and also seismic methods. These techniques can investigate the earth beneath its outer visible skin and in prospecting, and unravelling subsurface features.

Geophysics lets us 'see' below the surface. The first oil wells were drilled in areas where oil seeped to the surface, indicating that there was more to be found below the surface, but a seeping oil supply is a depleted oil supply, so seismic methods are commonly used to locate 'oil traps' in buried sediments, structures which catch oil and hold it in reservoirs below the surface.

An ore deposit which lies on the surface is generally an ore deposit which is partly weathered and dispersed. More importantly, an ore deposit may be close to the surface, but covered by soil or vegetation. Most ore bodies contain large amounts of metal-bearing ores which are more dense than the surrounding rocks, and these can be detected with a sensitive measurement of local gravity.

The basic law of gravitation says that every atom exerts a pull on every other atom, the force being proportional to the mass of the atom, and reducing with the inverse square of the distance between them. The 'pull of gravity' that we know from the earth is just the sum of all of those tiny pulls, acting on each of our atoms.

The variation that we call an anomaly is extremely slight, but it can still be measured, not by doing an absolute calculation of the value of acceleration due to gravity, but by comparing the pull from place to place. The variations being measured are typically one part in forty or fifty million, and absolute reliability of measurements at those levels is not always possible.

Under ideal conditions, modern gravimeters can detect fluctuations as small as one part in a hundred million, provided the gravimeter is operated on the ground. From the air, accuracies of better than a few mGal are unlikely—with normal gravity at 980,000 mGal, that means no better than one part in a quarter million.

The same ore bodies which cause gravity fluctuations may also produce small local variations in the magnetic field, and these can be detected by sensing devices, 'streamed' behind either fixed-wing aircraft or helicopters. The variations that are detected are known as 'anomalies'. Any unexpected values and results must have a cause, often a cause which can interest prospectors.

The most common form of geophysical searching involves using reflected noise, either from small charges of explosive, or from a mechanical source. The aim is to produce a point source of sound, and then to record the echoes heard at different points along a survey line, using sensitive microphones called geophones.

The signals detected by the geophones are then fed to a computer, which uses the data to build up a picture of what lies beneath the surface. The method only works because interfaces and boundaries between layers tend to reflect some of the sound that strikes them, while distance to a reflecting point is indicated by the time the echo takes to arrive.

Geochemistry goes hand-in-hand with geophysics, and it began with Robert Bunsen's chemical studies on igneous rock from Iceland and Armenia, as we saw in chapter 1. The term just means 'earth chemistry'. Most of the time, though, we can tell the rocks from each other by their appearance.

Geochemistry looks at the abundances of elements and their isotopes in the Earth, and the processes that affect their distribution. It also looks at chemical processes involved in the evolution of the earth. Commercial applications include geochemical prospecting, where the chemical analysis of soils, sediments, and stream waters is used to detect concealed ore deposits.

Unconformities and breaks in time.

An unconformity is the result of a gap in the geological record, where rocks have been eroded away, and later, new material has been added above. If the beds below the break are tilted, it may be referred to as an angular unconformity. Charles Lyell argued that some of the best geological evidence came from the unconformability of strata of different ages:

> Thus, for example, on the borders of Wales and Shropshire, we find the slaty beds of the ancient Silurian system curved and vertical, while the beds of the overlying carboniferous shale and sandstone are horizontal. All are agreed that in such a case the older set of strata had suffered great dislocation before the deposition of the newer or carboniferous beds, and that these last have never since been convulsed by any movements of excessive violence.
> —Charles Lyell, *The Principles of Geology*, 1853, 187.

It was a single unconformity that kick-started the whole science of geology, when James Hutton found an angular unconformity at Siccar Point in Scotland in 1788, a place where one set of horizontal sediments had been uplifted, folded and eroded, before other sediments were laid down over them. And here is Hutton's unconformity: the upper layer is the famous Devonian Old Red Sandstone, sitting unconformably on Silurian greywacke.

Hutton's Siccar Point Unconformity, Siccar Point, Berwickshire, Scotland. [Wikimedia Commons].

This led Hutton to believe that the earth was very old, but on theological grounds, he could not accept the idea that a divine Creator would make an earth which would wear out, so there had to be some mechanism of renewal. In his view, the earth was some sort of perpetual motion machine.

And so we got the uniformitarian principle, the idea that the forces now operating to change the earth's surface have always operated in the same way. There were no catastrophes, said Hutton, just slow and steady change.

> The result, therefore, of this physical inquiry is, that we find no vestige of beginning, no prospect of an end.
> —James Hutton, *Theory of the Earth*, 200.

So what does an unconformity look like? Here is another example, caught from a coach in Ecuador: at the top of the hill, the strata are horizontal, but half-way down, the beds are sloping down to the left at an angle of about 30°.

An unconformity, outside Quito, Ecuador.

As part of the work for this book, I set myself the task of locating points where the bottom of the Sydney Basin (Triassic and Permian rocks) sat unconformably on the underlying older rocks. I know three places where the boundary can be seen. The first is at Myrtle Beach, south of Sydney, if you can see the tilted metamorphic rocks below, and more or less horizontal rocks above.

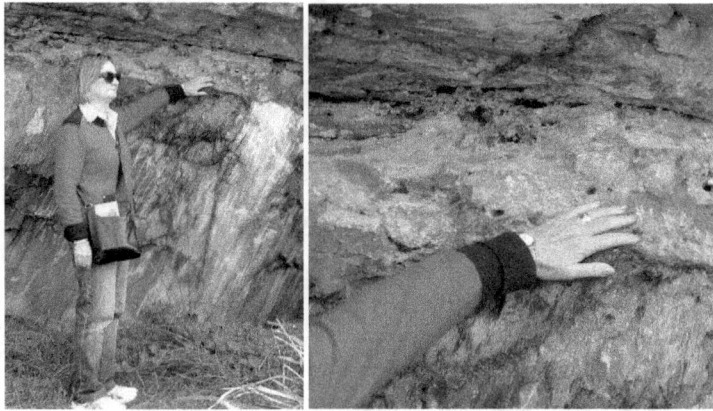

One hand spanning something over 100 million years, Myrtle Beach, NSW.

So how big is the gap? The Devonian era, according to the geological time scale, was 416 to 359 mya, while the Permian was 299 to 251 mya. So if the Devonian rocks beneath were laid down at the close of business on the last day of the Devonian, and the Permian rocks were laid down on the first morning of the Permian, the gap is 60 million years.

If we take the other extremes, the gap might be 165 million years: on average, it was probably a gap of somewhere between 80 and 140 million years. The other two places I know where you can reach the base of the Sydney basin are the subject of a tale with a moral.

First up, the Sydney Basin is the lump of geology that I live on top of, so I know it well. Centred on the coast at Sydney, the Sydney Basin reaches out west over the Blue Mountains, south past Ulladulla and north past Newcastle.

It is made up of Permian and Triassic rocks, lying unconformably on top of other rocks. There are scattered dykes, the occasional flow. and at least three volcanic necks that I can recall—and I think there are others I never knew about.

Once upon a time, just after the Carboniferous, rocks from the Devonian era which had been cooked out of recognition and squished way down, had become what we call metamorphic rocks. They were also folded or tilted, but by then, any Carboniferous rocks were eroded away and the Devonian rocks surfaced.

From either folding or tilting, the old rocks were all at an angle, and they got eroded back to a jagged landscape that sank down in the Permian, far beneath the sea. This happens in geology, so just live with it. At Myrtle Beach, the sediments that covered the old rocks were fairly fine, but in what is now the Budawang Ranges, inland from the Nowra-Ulladulla coast, a big flood must have pushed a huge supply of cobbles, mud and sand out over the newly submerged old rocks.

Remember that the original surface of the eroded Devonian rocks was a jagged one. I forgot that, and that was the seed of the surprise ending that came only after a lot of slogging. First, I made a one-day trip with my sons, about 15 or 16 km each way, but we had to stop, short of our target.

As a family, we have been going into the area for 40 years, but if you do the sums, I am necessarily slowing down. Unable to make it in and out in one day, we retreated in good order, getting out before it got dark. A few months on, we made a two-day trip, equipped to camp and walk out the next day.

On the way in, the track is familiar enough that we greet favourite trees, and favourite stopping-and-resting points, one of which is Tinderry Lookout. We were walking on Devonian metamorphics going uphill to the lookout, which is the first bit of serious Permian conglomerate that you see when walking in, but I still didn't wake up: perhaps you can see what we must have passed.

As I sat at the lookout, it struck me that somewhere on the way up to Tinderry Lookout, we must have crossed over the boundary, and by leaving our gear and walking slowly back down the hill, we just about located it, but it was buried, and I would never get a satisfactory photo of the sort people get at Siccar Point.

So we went on in, and made it to the camping cave, where I was very happy with the results: perhaps it wasn't Siccar Point, but it worked fairly well.

The top part is Permian conglomerate, the rest is Devonian metamorphics, tilting about 20° to the left.

And the stinger? On the way out, we stopped again at Tinderry Lookout. We had lunch, played with a skink that came begging, and then I saw **It**. *We were standing right on top of the unconformity!*

If you look very closely (or if you take this with you when you go there), you can see ribs of lichen-covered metamorphic rock on the ground.

So we had reached our goal, no sweat, no slog, no trouble! When the floods came, back in the Permian, the rubble and junk that later became conglomerate flowed into the deepest parts, levelling the surface and then filling in upwards, and I had NEVER noticed it! To be a good geologist, *you have to know how to observe*! Apparently, I'm still working on that skill…

Just over that drop, lies the unconformity I was seeking. I will get to it, one day…

The pendulum and the real shape of the earth.

In the 1600s duplicating and distributing standards for length and mass were a problem, but Christiaan Huygens thought he could solve the length problem.

He said that wherever you went, you could set up a pendulum, adjust the length so it had a period of one second, and get a length standard to refer to.

His scheme might have worked, and the value, around 99.4 cm, would have been near enough to the metre or the yard, but there was a snag: the value of g was found to vary from place to place—and sorting that out had some surprising ramifications.

In 1672 Jean Richer reported that the period of a pendulum varied with latitude, and Isaac Newton said that Richer's variation in the period of the pendulum was due to an equatorial bulge, and that offended the French. By this time, nobody thought the world was a perfect sphere any more, but France and England disagreed on the real shape, and national honour was at stake.

Newton said the Earth was an oblate (squashed) spheroid, a bit like a pumpkin. If this were so, argued Newton, something called the precession of the equinoxes could be explained. Against that, French scientists had taken some sloppy measurements, getting results which suggested an Earth more like an on-end watermelon than a pumpkin.

Richer's pendulum clock had been accurate in Paris but by then, astronomers had a very reliable 'clock', the way the four main moons of Jupiter disappeared behind the planet, or passed in front of it. So Richer knew that his clock lost two and a half minutes each day at Cayenne in French Guiana, closer to the equator. Clearly, more data were needed, and scientists were rushed to different places, mainly in South America, a mere 63 years after Richer's observations. Newton had died in 1727, but the French still wanted to show up that insolent upstart!

The *Académie française*, the French Academy, France's main scientific body, set out to determine the truth of the matter by experiment, measuring a degree of latitude in Lapland and in Central America. We will ignore the Lapland group for now, but the scientists who went to South America included Pierre Bouguer and Charles La Condamine.

Pierre Bouguer's name lives on today because geophysicists talk about *Bouguer anomalies*, which commemorates his pioneering work in the Americas. As we have seen, when the acceleration due to gravity is measured *very* accurately, small local fluctuations can indicate equally local deposits of high or low density mineral ores—these fluctuations are now called Bouguer anomalies.

Bouguer spent a much of his life studying gravitational effects, and in 1740, he tried to estimate the value of G, the universal gravitational constant, using a mountain as an attracting mass. That method can only be as accurate as the information the enquirer has about the interior of the mountain, and there were other problems which he could not have known about. We will ignore those here, but the key word is isostasy. Look around!

The aim of the French workers in Central and South America between 1735 and 1743 was to measure the length of an arc of one degree of latitude at the Equator. This was to be compared with the Lapland, close-to-polar, degree. Any difference in the lengths of the two degrees would reveal whether France or England was right about the shape of the planet.

Here, from the *Memoirs* of the French *Académie des Sciences*, is part of a letter from Bouguer to René de Réaumur in Paris in 1735, followed in the second paragraph by part of Bouguer's 1749 report of his findings. The main measurements have been converted to their modern equivalents. The unconverted "line" is 1/4 of a barleycorn, a twelfth of an inch, or about 2 millimetres.

> I have made here [in San Domingo] a simple pendulum of steel which I have made as invariant as possible. It has a bob of [12 kilograms], about [12 centimetres] in diameter and [3 centimetres] deep. To keep it swinging true, I have put on the rod a crossbar of iron to serve as an axis, at right-angles to the rod. The instrument is mounted on a tempered steel knife-edge on two steel springs. These two springs are mounted on a copper plate in which there is a hole for the rod. The plate rests on a stool [1.5 metres] high, and is levelled by three screws…
>
> We used the barometer that we set up to study the balance between the weight of the mercury and the air in all the accessible parts of the atmosphere. We saw how many feet we had to rise or descend to make the mercury change height by one line. It is then necessary to find the specific weight of air that balances other bodies. In this way, I have found by comparison with copper that on the top of Pichincha, there is a loss from unity of 1/11,000. Now it follows that the weight of my simple pendulum also loses 1/11,000 part of its weight. This loss produces a similar reduction in the restoring force, and naturally, I found the pendulum to be slow by 1/11,000. To correct this loss, it was necessary to adjust the pendulum's length by 4/100 of a line…

Translation of the translation: Bouguer had an accurate pendulum, mounted on a wooden stand (the "stool") and it was adjustable. He used a barometer as a way of measuring altitude. By timing the pendulum, he could get a measure of g at different heights above sea level.

The degree-measuring expeditions proved Newton correct, but one of the more lasting effects came from La Condamine's explorations while he was there, he travelled over a large part of South America, and then went 5000 km down the Amazon.

When he returned to Europe, La Condamine brought with him what the locals called *cauchu*, and the French still call *caoutchouc*. Thanks to Joseph Priestley, we still call it "rubber", because it can be used to rub out pencil marks, and what is called an eraser in some English-speaking countries is still a rubber in others.

Inconstant gravity.

The gravitational pull of the Earth varies from place to place. In general, it is smallest at the equator, and it is greater at sea level than at a higher altitude. Another factor which plays a part is the mean density of the rocks closest to the point where the acceleration due to gravity is being measured.

Place	latitude	elevation	g
Equator:	0°	0 metres	9.780 ms^{-2}
Pike's Peak:	38° 50'	4300 metres	9.789 ms^{-2}
Denver:	39° 45'	1650 metres	9.796 ms^{-2}
San Francisco:	37° 35'	100 metres	9.800 ms^{-2}
New York:	40° 45'	0 metres	9.803 ms^{-2}
North Pole:	90°	0 metres	9.832 ms^{-2}

How g varies with latitude and altitude.

In the list of g values above, the only one I could find, the American locations are all at nearly the same latitude, while varying in altitude, the other locations have values given for sea level, to allow easier and more interesting comparisons. All are actual measurements, with the exception of the North Pole value, which has been calculated from known data.

The total pull is the sum of all of the small pulls, and close to a large mountain range, the principle of isostasy ensures that the rock beneath the surface is less dense. In addition, the mass of rock above the Earth's surface in the nearby mountain actually produces a measurable upward component in the force due to gravity.

The effects of gravity had been well understood, ever since the time of the ancient Greeks, even if the nature and implications of gravity were less obvious. Until the 19th century, everybody 'knew' that everything fell to the centre of the Earth, that the plumb bob pointed straight down to the planet's centre.

It is one thing to know what gravity does, but it was altogether another thing to know why gravity does what it does, and a great deal more challenging to calculate gravity's effects. When Dan Brown asserted in *Angels and Demons* that at 60,000 feet, about 18.3 km, gravity is down to 70% of what it is at sea level, I had just enough background to know he was wrong. I had done a similar sum before.

I went out at Woomera a year or two before I read Brown's worthless scribble. I was there to cover a scramjet launch, a test engine that was carried to 320 km before dropping back to Earth. As I drove north from Port Augusta, it was a long haul on a road with giant road trains, kangaroos and emus, and I needed something to keep me alert. I did some back-of-the-mental-envelope calculations to see what the gravitational force would be at the top of its flight.

Over about 55 km of dead-straight road, I came up with the value 90.7% of ground-normal for gravity at 320 km, and I double-checked it. So I feel your pain, dear reader, in confronting such abstruse matters, but this is an area we must be on top of. That and not ending a sentence with a preposition, but talking of pain, would you believe that scientists call the pull of gravity a *weak force*?

If you had just stepped off a high bridge, you might feel the pain and say "those scientists are nuts!", but it takes an extremely heavy planet to deliver that much pull. The same bridge you have just stepped off is also heavy, but you don't even feel the pull that it is applying to you, because its mass is too small: *gravity is indeed a weak force.* Now here are the values you need to evaluate Dan Brown's claim: his value for 18 km, 70% of ground-normal, is only reached at 1250 km!

Km from surface	Acceleration due to gravity (ms-2)	Percentage of earth surface g
0	9.8	100 %
18	9.74510669	99.44%
100	9.500781065	96.95%
200	9.215059688	94.03%
320	8.888888889	90.70%
1250	6.859037122	70%
2500	5.067642974	51.71%
5000	3.088704217	31.52%
6400	2.45	25%
385000 (moon)	0.00262026	0.026%
150 million (sun)	0.000000178	0.00000018%

The measure of acceleration is metres per second per second (usually written as ms^{-2}), and an acceleration of 9.8 ms^{-2} just means that if you are falling at 100 metres per second, for each second after that, you add 9.8 metres per second to your velocity. That is, you accelerate by (9.8 metres per second) each second.

Physicists say the acceleration due to gravity varies in accordance with the inverse square law. While this sounds complicated, it's a shorthand way of saying that when you are twice as far from the centre of gravity as before, the gravitational pull will be just a quarter of what it was originally.

So that's clear, but what's the centre of gravity? That's hard to define in words, so the next section will look at this strange idea and how to use it.

The centre of gravity.

This is the single point in any body with mass which may be regarded as having the entire mass concentrated there. When a body is hanging up, a plumb line from the point of suspension always passes through the centre of gravity, so the intersection of two or three such lines can be used to locate the centre of gravity.

A force along a line which passes through the centre of gravity will move the body without making it rotate (or change its rotation), but it should be noted that some bodies, like horseshoe shapes and hollow spheres, may have a centre of gravity which does not lie within the material of the body itself.

For a sphere, the centre of gravity is just the centre, the middle. When heavy bodies orbit each other, they move around the shared centre of gravity, and this is how we find exoplanets, the planets that don't orbit our Sun.

Imagine two skaters with their hands clasped, swinging around on the ice. You don't see one circling the other, you see them circling a central point, where their hands meet: they both orbit around their common centre of gravity.

If you replace one skater with a Sumo wrestler and the other with a small child, the rotation point moves. The wrestler would still wobble a bit, but it would look as though the child was almost in orbit around him. With a heavier second skater, the wobble effect would be reduced.

When a planet orbits a star, they both rotate around the centre of gravity, and the star seems to wobble like the Sumo wrestler. When the planet is moving away from us, the star is coming our way, and *vice versa*. That means we can look for Doppler shifts that indicate the presence of a planet, and even learn its orbital period and distance.

Just 15 light years away, a rocky world, 7.5 Earth masses and about twice the Earth's diameter, whizzes around Gliese 876 (or GJ 876), a star just like our

Sun. It does so about 3 million km away from the star, which means it is rather hotter than our planet, to say the least of it. You could bake a roast on it (even though GJ 876 is in the constellation Aquarius!).

Our sort of life would be impossible there. The years take just 1.94 of our Earth days, but maybe life in the fast (p)lane(t) is different! Then again, there may be other smaller and slower planets still to be found around GJ 876, but it is the physics of the centre of gravity that tell us what is what.

It was a silly geographical error which drew Christopher Columbus to discover the Americas, and it was an equally silly error which was one reason for James Cook to visit Australia and New Zealand, in search of a Great South Land, at least partly because of the 'counterpoise argument', which said that there had to be a large antipodean land mass to balance the weight of the northern continents.

Alexander Dalrymple believed there needed to be a large mass of land. It was, he said, "…wanting on the South of the Equator to counterpoise the land to the North, and to maintain the equilibrium necessary for the Earth's motion". Perhaps, if anybody had stopped to think what the earth weighed, Cook might not have searched the South Pacific and found Australia.

Unlike an unbalanced fly wheel, the distribution of the mass of surface land is not a serious issue in the balance of the globe, which merely rotates around its centre of gravity, without being in any way influenced by 'imbalances'. Only mechanical devices that are spinning on axles will wobble. Any irregular body, rotating freely in space, can turn freely.

Leaving out hurdling and diving, there are four main forms of competitive jumping: the high jump, the long (or broad) jump, the hop-step-and-jump (which gained dignity by becoming the triple jump), and the pole vault.

The high jump involves raising your centre of gravity as high as possible while, at the same time, getting all of your body parts over a bar that hangs precariously on two uprights. Success comes from running as hard as you can at the jump area, and then converting your forward energy into a vertical leap.

Your centre of gravity is a little over halfway up your height, somewhere not far from your navel. If you are two metres tall, and clear the bar at two metres, you have succeeded in raising your centre of gravity by about 90 cm. Now by a neat bit of physics that we will come to in the next chapter, your upward speed must be the same as an object falling to the ground from 90 cm, so your upward speed as you took off was 4.2 metres/sec, just over 15 km/h, a third of the maximum speed that a sprinter can achieve. The current world record of 2.45 metres is equal to about 5.2 metres/sec of vertical speed at take-off.

A top pole vaulter can clear 6 metres, which is equivalent to almost 10 metres/sec, or 36 km/h, close to the world record for sprinters. The catch is that part of the energy comes from the run-up, but there is more as the runner pushes forward when the pole digs in, and more energy from the jumper's arms and shoulders gaining leverage on the pole.

For male long jumpers, the approximate velocity v for a jump of s metres is $(s + 2.23)/0.95$ m/s, while for females, it is $(s + 2.81)/0.99$ m/s. Jesse Owens' 1935 record of 8.13 metres implies an improbable 11 metres/sec, so use this formula with care!

Most people know that gravity on the Moon is about $1/6$th of that on Earth, so they assume that Javier Sotomayor, who cleared the high jump bar at 2.45 metres would be able to clear 14.7 metres on the Moon, assuming he was in an enclosure and did not need a space suit.

If we leave out air friction which would have to be there if the humans were breathing, the athletes who threw things would probably get close to doing six times better than on earth, but there is a catch in non-throwing events: running fast needs gravity, and this might even affect some of the throwing events.

There is no easy way to do the analysis, but a sprinter who takes off out of the blocks needs to get a grip on the ground, and under low gravity, the runner would probably lose contact with the ground, and be unable to keep those legs pushing forward. On top of that, some part of the movement of the arms and legs relies on a pendulum effect, and the period of any pendulum depends on the gravitational forces acting on it.

In other words, there are good reasons to expect that running on the Moon would be harder than it is on earth. Let us set that aside for a moment. Sotomayor's 2.45 metres implies a vertical speed of 5.2 metres/sec. If we take the acceleration due to gravity on the lunar surface as 1.62 ms^{-2}, we find that Sotomayor would raise his centre of gravity 8.35 metres, which would see him just clear around 9.5 metres, more than 5 metres short of the predicted height.

That would be fairly spectacular, but keep in mind that we made the dubious assumption that Sotomayor could run as fast on the Moon as on Earth. It might be safer not to bet on the high jump at any future Lunar Olympics!

Blue whales can leap completely out of the water, which probably involves raising their centre of gravity by 8 to 10 metres. Using the principle that speed-up = speed-down, it is easy to show that their speed when they fall back is equal to their speed as they started to emerge, and a 9 metre fall means a speed of 50 km/h, near enough, but we are leaving earth science behind.

13. Why dropping rocks on toes hurts.

What doth gravity out of his bed at midnight?
— William Shakespeare (1564 – 1616), *Henry IV Part 1*, II, iv

Yet out of pumps grew the discussions about Nature's abhorrence of a vacuum, and then it was discovered that Nature does not abhor a vacuum, but that air has weight; and that notion paved the way for the doctrine that all matter has weight, and that the force which produces weight is co-extensive with the universe — in short, to the theory of universal gravitation and endless force.
— Thomas Henry Huxley (1825 – 1895), *On the Advisableness of Improving Natural Knowledge*, 1866.

The airplane stays up because it doesn't have time to fall.
—Orville Wright (1871 – 1948), explaining flight (in a way which better explains the orbit of the moon).

But I, Simplicio, who have made the test can assure you that a cannon ball weighing one or two hundred pounds, or even more, will not reach the ground by as much as a span ahead of a musket ball weighing half a pound, provided both are dropped from a height of 200 cubits.
— Galileo Galilei (1564 – 1642), *Dialogues Concerning Two New Sciences, First Day*, Dover, 1954, 62.

You can drop a mouse down a thousand-foot mine shaft; and, on arriving at the bottom, it gets a slight shock and walks away, provided that the ground is fairly soft. A rat is killed, a man is broken, a horse splashes.
— J.B.S. Haldane (1892 – 1964), 'On Being the Right Size', from *Possible Worlds*.

If you fall from a high tower, you fall quicker and quicker; a judicious selection of a tower will ensure any rate of speed.
— Stephen Leacock (1869 – 1944), *Storybook Lapses* (1910)

Gravity and gravitation.

Most people would say 'gravity' was discovered by Sir Isaac Newton, but gravity was known long before his time. Getting a bit technical, Newton was considering the orbit of the moon, and he showed that the force of gravity must diminish in accordance with the inverse square law.

Under that law, doubling the distance from any heavy body reduces the attractive force to a quarter, but note that the gravitational pull of a sphere has to be treated as a point-mass at the centre of the sphere.

In practical terms, we stand on the surface of the earth, some 6400 km from the centre: if we are raised to a height of 6400 km above the surface, the pull of gravity on us would be reduced to a quarter of what it is now. The attraction of gravity on the surface of the moon is about one-sixth of what it is on the surface of the earth, simply because the moon has a smaller mass than the earth.

Interestingly, most people will say wrongly that a pencil, released by an astronaut on the lunar surface, will either float where it is or drift away. In fact, the pencil will obey the laws of gravitation, and fall towards the moon. If you ask people to explain their answers, they will usually say that there is no gravity in space, or that the vacuum on the moon, the gravitational pull of the earth, or some other cause has neutralised the moon's gravity. If you ask why astronauts don't float away, people often say this is because astronauts wear heavy boots!

As we have seen, a high jumper on the moon cannot leap six times as high as the same athlete on earth. Let's stay with physics of falling, then. When I was a schoolboy, we physics people asked: "if you jumped from the Empire State Building, would you penetrate the roadway or splatter?

Lacking sufficient data, my friends and I concluded that you would splatter as you went through (well, we *were* schoolboys *and* physics students! Think of us as engineers with a charisma bypass…)

We can use the simple equation $v^2 = 2as$ here, because the initial velocity is zero. Young readers, please note that this is one of those considerations that physicists call a thought experiment or *Gedankenexperiment* if you are German. Whatever your origins, do *not* try this at home!

As you can see from the table below, it takes 2.5 seconds to fall the first ten floors, but just over a tenth of that time to cover the last ten. For an elegant ending, you should shout "so far, so good," as you pass the third floor. You will need to speak fast, but sadly, you will never know if your listeners got the joke.

Storeys	Metres	ms⁻¹	km/hr	Secs
1	3	7.7	27.8	0.8
2	6	10.8	39.0	1.1
3	9	13.3	47.8	1.4
4	12	15.3	55.2	1.6
5	15	17.1	61.7	1.7
10	30	24.2	87.3	2.5
15	45	29.7	106.9	3.0
20	60	34.3	123.5	3.5
30	90	42.0	151.2	4.3
40	120	48.5	174.6	4.9
50	150	54.2	195.2	5.5
60	180	59.4	213.8	6.1
70	210	64.2	231.0	6.5
80	240	68.6	246.9	7.0
90	270	72.7	261.9	7.4
100	300	76.7	276.1	7.8
110	330	80.4	289.5	8.2
120	360	84.0	302.4	8.6
130	390	87.4	314.7	8.9
140	420	90.7	326.6	9.3
150	450	93.9	338.1	9.6

Falling off a building (not recommended).

The idea of acceleration came through Newton, who got it from Galileo Galilei, and then passed it on. Galileo was probably the first to realise that it would be easier to time objects sliding gently down a slope than it was to time objects in free fall. This, together with the primitive timing devices which he and his colleagues made, allowed Galileo to see clearly, for the first time, what acceleration really was.

Galileo used slow sliders because he had no accurate timing device. If the shortest unit of time that you can estimate is a second, this makes it very hard to observe and measure any acceleration.

He recognised that the final speed at the bottom would be the same, whether it was achieved directly through a simple drop, or indirectly after running down a long ramp, the sort of ramp that physicists call an inclined plane.

Of course, every discovery has to have a use in war, if you want funding. We hear inspiring tales of scientists, asked by rulers what use their discovery is, who reply bravely, "what use is a baby?" or "one day, you may be able to tax it", but more often, they were ready with a proposed military use. This is why Galileo suggested that his telescope would be useful in identifying enemy ships while they were still a long way off the port that they were coming to attack.

So in the Renaissance, when physicists were able to explain the flight of a cannon-ball, the way was set to produce tables which would tell ignorant soldiers the angle and gunpowder load needed to hit a target, and they were on a winner, funding-wise. 'Gunnery tables' were one of the needs which drove the development of computers during World War II.

Newton had another equation that described the gravitational force between two masses, M_1 and M_2, when the distance between their centres of gravity was r, and it went like this:

$$F \propto (M_1 \times M_2)/r^2$$

Here, the sideways 8 means "is proportional to", which means the two values are in a standard proportion. So by choosing the right value for G ("the Universal Gravitational Constant"), we can write:

$F = (G \times M_1 \times M_2)/r^2$ But what was the value of G? That's coming next!

The Universal Gravitational Constant.

> G is one of those handy numbers that physicists know and use as part of their passwords or their PIN for bank accounts. Others include the first few digits of pi, e, or the square root of 3, but G is their favourite, even though G had been calculated by a geologist and parson, and not a physicist at all!

> You think I'm joking? I am giving away nothing at all about my passwords and PINs if I mention that my mobile phone number can be calculated (by approved initiates {=geeks} only) in two parts, from the square root of 2 added to the Euler identity ($e^{i\pi}$), followed by e to six significant figures. Scientists are like that.

There is a passage that always lurks in my mind when I think of the torsion pendulum, but it will only make sense if you have read Huxley's book, and know what happened to John Savage at the end.

> Slowly, very slowly, like two unhurried compass needles, the feet turned towards the right; north, north-east, east, south-east, south, south-south-west; then paused, and after a few seconds, turned as unhurriedly back towards the left.
> —Aldous Huxley, *Brave New World*.

Scientists needed to measure what the value of *G* was, but remember that gravity is a weak force, so it wouldn't be easy, and it would need a very clever design. Enter a man who, in many ways, made Newton look fairly normal, because this chap made many discoveries, but could rarely be bothered to publish them.

This was Henry Cavendish, though the idea came from the Reverend John Michell (pronounced Mitchell), who was also the first person to predict that black holes might exist. He is little-known, and was, in fact, a geologist and parson, not a physicist at all. Science was like that, back then.

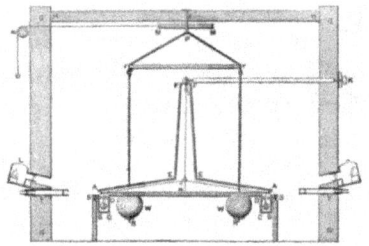

The Michell/Cavendish apparatus.

Michell wanted to study gravitational effects, and constructed a torsion pendulum in about 1783, but he died in 1793, without having used it. His friend Cavendish acquired the device and rebuilt it. A simple swinging-weight pendulum can be used to estimate the Earth's gravitational force by knowing the period of the swing and the length of the pendulum, but this device could weigh the planet.

Charles Coulomb used a torsion pendulum to measure tiny electrical forces and establish Coulomb's Law in France, at about the same time, but from Cavendish's notes, we know Michell thought of the idea first. The torsion pendulum was one of those ideas that were obvious to physicists back then.

The gadget measured the deflection of a beam holding one pair of lead balls when two other lead balls were brought near them, gauging extremely small gravitational forces. The 'wooden arm', by the way, was a truss structure. Here is how Cavendish described it:

> The apparatus is very simple; it consists of a wooden arm, 6 feet long, made so as to unite great strength with little weight. This arm is suspended in an horizontal position, by a slender wire 40 inches long, and to each extremity is hung a leaden ball, about 2 inches in diameter; and the whole is inclosed in a narrow wooden case, to defend it from the wind.
>
> As no more force is required to make this arm turn round on its centre than what is necessary to twist the suspending wire, it is plain, that if the wire is sufficiently slender, the most minute force, such as the attraction of a leaden weight a few inches in diameter, will be sufficient to draw the arm sensibly aside. The weights which Mr. Michell intended to use were 8 inches in diameter.

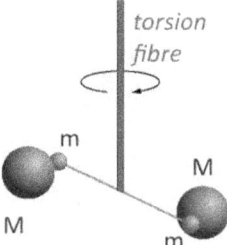

How the torsion pendulum worked: two larger masses (M) attracted the small balls (m).

The essential point is that the force of gravity acts between any two bodies, no matter how small they are. This force is proportional to the products of their mass, divided by the square of the distance between them. We know that, because Newton said it was so.

What Cavendish did next was outrageously brilliant. Once he had a value for G, he knew the exact force of gravity operating on a weight, and how much it weighed, and in the late 1790s, he had a good idea of the size of the Earth.

Now he could calculate the missing value, the mass of the planet and so measure the average density of the Earth, obtaining a value of 5.48, which was quite close to the modern accepted value of 5.52. Gravity still had a few tricks up its sleeve, even then. When surveyors started working close to the foothills of large mountain ranges, with high levels of accuracy, they realised that their plumb bob was no longer pointing straight up and down.

Instead, it was pointing slightly towards the mountains. At Kalianna, near the Himalayas, for example, the discrepancy was enough to produce a difference in the calculated latitude of 5.23 seconds, a distance of about 150 metres. It was immediately obvious that being close to massive mountains, gravity would pull a little bit sideways.

A theoretical calculation showed that this distance should have been about three times as great as it was, assuming that the Himalayas lay on an otherwise uniform crust.

We now explain this difference between theory and reality by referring to the geological principle of isostasy. In brief, this principle says that mountain ranges are "floating in the crust", and the piece of mountain we see above the surface is like the tip of an iceberg, only showing up because the whole mountain block, including the part underground, is less dense than the rock in which it sits—just as the underwater ice of the iceberg is less dense than seawater.

And the actual value of *G*? For most purposes, it is taken as either 6.674×10^{-11} m^{3}kg^{-1}s^{-2} or 6.674×10^{-11} N(m/kg)^2. Don't worry too much about the units: just stick with the value.

Gravity and golf in space.

In 1971, astronaut Alan Shepard played the first-ever golf stroke in space when he struck a ball out across the lunar surface. He played with an improvised club: a Wilson six-iron head attached to a lunar sample scoop handle. He claimed later that the second of two balls went "miles and miles".

In 2006, cosmonaut Mikhail Tyurin played a shot of sorts from the International Space Station. He did it one-handed, something that would hardly be approved of on most courses. Worse, US astronaut Michael Lopez-Alegria held onto Tyurin's feet while he made the stroke. This stopped him accidentally flying off into space, but it would never be accepted at the Royal and Ancient!

The ball weighed far less than a standard ball, just 3 grams, and Tyurin hit it with a gold-plated six-iron, as part of a stunt to promote a Canadian golfing equipment company. He shanked the shot, but the ball was still expected to burn up in the atmosphere after several days. With that sort of time frame, even a gentle tap would result in what the company claimed was "the longest drive".

There are many who have claimed this title, though quite a few of the alleged records were achieved far from the golf course: strokes have been played along airstrips, from hills overlooking frozen rivers, and in other conditions that would artificially enhance the "carry" of the ball.

The flight of a ball on the moon or from a space station would differ remarkably from a ball hit on a golf course. The stroke itself uses muscle power to accelerate the club head towards the ball, but the force of gravity also plays a small part. The lower gravity on the moon and the "lack of gravity" in space (there *is* gravity in space, you just can't notice it) play a bigger part.

Then there is the influence of the air. From the space station, the ball would have followed a complex curve, shaped first by the Earth's gravity and then by drag from the wisps of atmosphere that it ran into. On the moon, the ball would describe a parabola, moving forward as fast when it touches down as it moved when it left the tee.

On Earth, the ball climbs steadily, and sometimes even begins to climb more steeply, before it slows, and drops more or less straight down to the fairway. This strange path is predictable from the way the ball starts out. With a perfect on-centre hit, the ball climbs at a shallow angle, with a backspin of as much as 3000 rpm and travelling at around 70 metres/second.

By the time it falls to the grass, the ball's forward motion will be less than a third of that. Not all of it has been wasted, though, because some of the energy carried by the ball was used to keep it aloft so it could move further forward.

The secret is an aerodynamic effect called Magnus lift, and this is why there are dimples on a golf ball. They help form a turbulent layer underneath the ball, and that keeps it floating along. This was not something the physicists thought up: it all began when canny golfers noticed that their older and more battered balls went further. Speed was a key factor, but a dimpled ball was better!

Still, for most aspects of golf, speed is all-important, but when it comes to putting on the green, there is a severe speed limit. On Earth, if your ball is travelling faster than 4.6 km/h, and it hits the hole squarely, it will cross the cup before it drops far enough to be stopped by the opposite lip. If you want your ball to go in, rather than hopping up and going over, that is one speed you *must not* exceed. On the moon, the limit would be a great deal lower.

An asteroid named Eros.

The first asteroid known to humans was discovered on January 1, 1801, but the best-studied one, Eros, was discovered almost in the middle of the 'asteroid age', in 1898. It was the first discovered asteroid that crossed the orbit of Mars, and it was rapidly seen to be irregularly shaped, based on fluctuations in its brightness as it rotated once every 5.27 hours. Eros moves to as close as 1.13 AU and as far as 1.73 AU from the sun, in an orbit that takes 1.76 years.

As we have seen, golf has been played in space, both on the moon and from a makeshift tee on the International Space Station. In each case, it was just a single drive, with no attempt to complete the hole, but it occurred to me to wonder what prospect there would be of playing a round of golf on the asteroid 433 Eros, and I sat down to look at the physics of the case.

I can date this pondering with some precision: it came a few days after January 10, 1999, which was when the NEAR Shoemaker spacecraft passed close to the asteroid, taking photos of the surface as it went. The images were fuzzy, but the asteroid's dimensions were set down: it was now officially 33 x 13 x 13 km, as opposed to a previous estimate of 40.5 x 14.5 x 14 km. We knew the asteroid rotates once every 5.27 hours and had a specific gravity (density) of 2.7

Note the long and short dimensions: 433 Eros is built like a cylinder, a can of drink, with one long dimension, and that will become important.

There was a ridge running 20 km along the asteroid, and this and its high specific gravity suggested that Eros was a homogeneous body, not a collection of rubble as some scientists had thought. Significantly, the surface of Eros has many craters. The two largest craters are 8.5 and 6.5 km in diameter. That was when I started my golf ponderings—and please keep in mind the shape of the asteroid.

For a competent physicist, the calculation of gravitational attraction is a doddle, if you know the distance from the centre of gravity of the asteroid to the place where the measurement is taken.

I knew a competent physicist, who gave me the figures and this is where things get seriously weird. For a start, unlike me, he was a golfer, and saw right away that any player on the surface of Eros would probably fly away.

If you stood in the middle, you would be just 6.5 km from the asteroid's centre of gravity, but at either end, you would be 16.5 km away, and the inverse square law comes into play. There is a major difference in the gravity you would feel, depending on where you stood. Walking to the end of Eros would, in fact, feel like walking uphill, because the physics of a drink-can body are not those of a tennis-ball body. You have to work to get away from the centre of gravity.

The acceleration due to gravity on the surface of Eros varies depending on where it is measured, ranging from 2.3 to 5.5 mm/s^2 (that is millimetres, not metres!) some 2000 to 4000 times smaller than on Earth.

A person who weighs 70 kg on Earth would weigh from 16 to 39 grams on Eros, depending on where they were. Eros has an escape velocity that ranges from 3.1 to 17.2 m/s, again depending on the point on the asteroid where it is measured, which would allow a golf ball hit from its surface to leave forever. That is a drive between 11 and 61 km/h, a range well within the ability of even the rawest beginner. Throw a club away in anger, and you may never see it again!

Well, perhaps you could try gentler drives, but then comes the problem of holing out. As we saw earlier, a ball on a green on Earth, travelling faster than 4.6 km/h (and less on the moon), will bounce right over the standard hole, but to have a ball drop into the cup on 433 Eros, either the putt must be exceedingly slow, or the cup so wide as to make the game pointless. So golf simply cannot be played on 433 Eros!

Footnote: The low gravity challenged scientists getting NEAR Shoemaker into orbit around the satellite, as the craft needed to have a relative speed of less than just 6 metres per second, 22 km/h. Sometimes slow is good!

Footnote 2: Your author partly financed his undergraduate studies by caddying, but he is not a golfer, because the game is frustrating: even a good player takes 72 strokes to hit the hole 18 times, meaning a good golfer has a failure rate of 75%. Note that the word 'golf' is 'flog' backwards.

Flying away forever.

I mentioned escape velocity in the last section but did not explain it. This is the speed needed by an object leaving the surface of a body such as a planet forever, so that the object can just escape the gravitational pull of the larger body. The escape velocity at the surface of the Earth is about 11 km/s, while on the asteroid Eros, sports such as golf will never be played, because a reasonably driven golf ball would exceed the escape velocity of Eros.

For any planet, there is always an escape velocity, a speed that a bullet or other object must travel at if it is to escape the surface gravity of that planet forever. Jules Verne knew this. When he wrote *From the Earth to the Moon*, he proposed sending people to the moon in a sort of tin can fired from a gun. If anybody had ever tried this method, they would have ended up as tinned soup.

A long barrel would have made the acceleration more gentle, but even if the barrel could be as long as the Eiffel tower, the acceleration force would still be 3500 times as great as Earth's gravity, and making the barrel as high as the Empire State Building would only reduce the force to 3000 times that of Earth's gravity, usually referred to as g.

The most a human can take is about 10g, so the gun solution would never work for launching humans. Verne probably knew that it would not work, but he needed a transport method to hang his story on. He certainly knew what the Earth's escape velocity was:

> If the projectile kept indefinitely the initial speed of 12,000 yards a second, it would only take about nine hours to reach its destination; but as that initial velocity will go on decreasing, it will happen, everything calculated upon, that the projectile will take 300,000 seconds, or 83 hours and 20 minutes, to reach the point where the terrestrial and lunar

gravitations are equal, and from that point it will fall upon the moon in 50,000 seconds, or 13 hours, 53 minutes, and 20 seconds. It must, therefore, be hurled 97 hours, 13 minutes, and 20 seconds before the arrival of the moon at the point aimed at.

Verne was also clever enough to see that there was an advantage in a launch site near the equator, and placed his fictional gun in Florida, where Cape Canaveral is located today. At 28° north, this is not as good as the 4° north of the European Space Agency's Kourou site in French Guiana, but it beats the Russian Baikonur cosmodrome, which is at 47° north in Kazakhstan.

An equatorial launch means the rocket already has a speed to the east of about 1650 km/h, with respect to the Earth's centre, before the engines start. That means that the rocket needs to burn less fuel to get into orbit or to reach escape velocity.

It is not only planets that have escape velocities: if a spaceship is ever to escape our galaxy, it needs to build up a speed of about 1000 km/s, and more to get out of the Milky Way. Escaping our planet is about 1/90th of that at 11.2 km/s: our planet has much less mass, but we are much closer to it, and distance counts. Here are some comparisons:

To escape:	Speed (km/s)	Speed (km/h)
Moon's equator	2.4	8,640
Mercury's equator	4.4	15,840
Mars' equator	5	18,000
Venus' equator	10.4	37,440
Earth's equator	11.2	40,320
Uranus' equator	21.3	76,680
Neptune's equator	23.6	84,600
Saturn's equator	35.5	127,800
Jupiter's equator	59.5	214,200
Sun's equator	617.5	2,223,000
the Milky Way	1000	3,600,000

14. Oh dear, is that the time?

All of recorded human history is much less than 4.5 billion minutes. Yet, geologists claim that Earth formed half-a-million times longer ago than that. No one can easily fathom the meaning of "deep time." So how can we be sure such age estimates are correct? The answer lies in the testimony of the rocks.
— Robert M. Hazen, 'How Old is Earth, and How Do We Know?', *Evolution: Education and Outreach* (2010) **3**:198 – 205.

We must remember that measures were made for man and not man for measures.
— Isaac Asimov (1920 – 1992), *Of Time and Space and Other Things*, 1965.

The world and time had both one beginning. The world was made, not in time, but simultaneously with time.
— St. Augustine of Hippo (354 – 430), *The City of God*.

Come what come may,
Time and the hour runs through the roughest day.
— William Shakespeare (1564 – 1616) *Macbeth*, I, iii, 146-7.

The first grand discovery was time, the landscape of experience. Only by marking off months, weeks, and years, days and hours, minutes and seconds, would mankind be liberated from the cyclical monotony of nature.
— Daniel J Boorstin, *The Discoverers*, page 1

We can be certain that the radiation did not change appreciably during the last 500 million years; because during all this time life existed on earth, which means that the temperature of the earth during the whole period must have been very nearly what it is today. This temperature is determined by the sun's radiation.
— Hans Albrecht Bethe (1906 – 2005), *The Sky*, December 1940.

Compared with what we think of as long periods in our everyday calculations, there must have been enormous time and considerable variations in circumstances for nature to lead the organisation of animals to the degree of complexity and development that we see today.
— Chevalier de Lamarck (1744 – 1829), *Philosophie Zoologique*.

DUCHESS OF BERWICK: …dear Agatha and I are so much interested in Australia. It must be so pretty with all the dear little kangaroos flying about. Agatha has found it on the map. What a curious shape it is! Just like a large packing case. However, it is a very young country, isn't it?
HOPPER: Wasn't it made at the same time as the others, Duchess?
— Oscar Fingall O'Flahertie Wills Wilde (1854 – 1900), *Lady Windermere's Fan*.

A Geological time scale

Ediacaran period 635 to 542 mya;

Cambrian period 542 to 488 mya;

Ordovician period 488 to 444 mya;

Silurian period 444 to 416 mya;

> Devonian period 416 to 359 mya;
>
> Carboniferous period 359 to 299 mya;
>
> Permian period 299 to 251 mya;
>
> Triassic period 251 to 200 mya;
>
> Jurassic period 200 to 145 mya;
>
> Cretaceous period 145 to 65 mya;
>
> Tertiary period 65 to 2.6 mya;
>
> Quaternary period 2.6 mya to the present.

The 6000-year Earth and the geologists.

At the start of this book, in *Background to earth science*, we looked back to a time when the source of all knowledge was the Bible, and the ways in which it could be interpreted. From that, many 'scholars' said the planet was created 6000 years ago, but the study of geology called that dating into question. It didn't add up.

Hugh Miller, geologist and stern Scottish churchman, died in 1856. In 1857, his widow referred in a new edition of his *The Old Red Sandstone* to "infidels" among the geologists, so clearly the lines were being drawn on the Biblical age of the Earth, well before Darwin published *The Origin of Species* in 1859.

A few geologists joined Mrs Miller in her fundamentalist approach to the age of the Earth, but the professional geologists and the vast majority of trained scientists already accepted that life had been on Earth far longer than the literal 6000 years that could be read into a literal reading of the Old Testament.

Reading the rocks requires a far cleverer type of literacy. Fossils are curious things, and fossil experts are adept at detecting slight variations that reveal hidden secrets. Most fossils carry subtle clues in their shape, their form, where they lie, or what lies around them, but perceiving this only comes after looking at large numbers of fossils with a clever eye.

That sort of insight does not necessarily help explain how a fossil came to be where it was, but it is a start. After that, you are left with a choice between logical reasoning and inference, or supposition and wild fantasy. Many people, finding a conclusion they don't like, will denounce another's logical reasoning as crazy fantasy, or hail a colleague's wild surmise as pure gold. Those who do this can sometimes be anti-scientists of the worst sort, but they can also be scientists.

The mainstream geologists were annoyed by catastrophists, people who wanted all geology to have been produced in several major disasters like Noah's flood. The normal geologist's view is uniformitarian, meaning that conditions have been the same, uniform, over the eons, with geology caused by processes we can see today, with weathering, erosion, volcanoes and other ordinary events shaping the Earth.

Some of the modern opposition to asteroid theories that account for the "end of the dinosaurs" stems from this same visceral reaction to any suggestion that catastrophes shaped the Earth.

Rational geologists now tend to assume a sort of geological punctuated equilibrium, where normal conditions apply most of the time, with the occasional surprise. All the same, geological mavericks who stress tsunamis, asteroids and other catastrophes tend to be looked down on, even today.

Boucher de Perthes had been an extreme catastrophist, and he was an amateur at a time when geologists were becoming professional. It took the mainstream scientists a while to trust him, but he just kept on, digging interesting human-made tools from deep chalk deposits in France.

In the end, the scientists came around to his ideas (or his results, anyhow), thanks mainly to Charles Lyell, who visited Boucher de Perthes' excavations in 1859 and came away convinced that the tools were not only real, but offered strong evidence that humans were older than supposed.

Lyell presented a paper on the topic at the September meeting of the British Association and published *The Antiquity of Man* in 1863. After his visit, Lyell wrote "That the human race goes back to the time of the mammoth and rhinoceros (Siberian) and not a few other extinct mammals is perfectly clear..."

The dispute over flint tools in *The Times* in November was all about age. The tools lay far deeper than tombs which contained coins 2000 years old. The Biblical 6000 years was too short for the depths at which the flints occurred, unless you assume a change in conditions.

A massive flood like Noah's might explain the deep burial, but the geologists knew that chalk beds are formed slowly by tiny organisms, not by floods. "The discovery of these relics of a race which seems to have been of far greater antiquity than any that has been hitherto supposed to have inhabited our planet, involves many interesting and difficult questions," wrote T. W. Flower in *The Times* on 18 November.

A correspondent from Notting Hill mentioned London's new sewers in *The Times*, noting that "a half-mile cut at Shepherd's Bush, 35 feet deep, part of the new London sewers", was to be extended all over London. The writer wondered if the Geological Society could appoint competent people to visit these cuttings and take note of the strata revealed.

The debate showed the opponents, the nay-sayers, as anti-Micawbers. Dickens' Micawber spent his life hoping for something to turn up, these people spent their lives scrutinising the works of science, hoping for something to turn down. 'Senex' was quick to propose a wild fantasy to account for the flints, and just like today's creationists, he based it on a garbled and cherry-picked version of the works of the enemy. He wrote in *The Times* on December 5.

> Mr Darwin tells us of cliffs in Patagonia full of fossils. Suppose such a cliff face fallen down over an Indian burying ground, and a river afterwards diverging from its bed depositing a variety of detritus, or drift, including minerals from the Andes.

The debate was not straightforward, because physicists also questioned the geological view on the planet's age from their own scientific perspectives. The planet was warm below ground, they said, and must be cooling, which meant it used to be hotter, because there was no apparent source of continuing warmth.

If you went back far enough, they said, the planet would have been too hot for life, and that set a limit to the time life had existed. The answer to this paradox was that internal radioactivity has kept the planet warm for billions of years, but in 1859, radioactivity was unknown, and it is the decay of radioactive nuclei that keeps the planet's interior hot.

The flint tools argument might have been good training for the debate that would descend, like Noah's deluge, once Darwin's book came out—and it was released on 24 November. People just needed to read it first. Today, it remains good training for those trying to convince palaeo-conservatives that their love of coal will kill our grandchildren.

Remember, we are talking science here, and science never offers proof, just evidence, although Thomas Jefferson managed to prove from scientific data that Noah's flood could *not* have happened. All the same, when so many different classes of evidence point in the same direction, perhaps we should forgive those scientists claim to have "proof that evolution happened".

There remains an important question: which piece of evidence is more important? The general story we get is always the same, but sometimes the competing "stories" point in slightly different ways, and in the end, we may have to choose between looking at the shapes of bones, the blood chemistry, or the shape and numbers of the teeth. Personally, the blood chemistry and DNA evidence win my allegiance, every time.

Evolution is well demonstrated over geological time by the sequence of organisms preserved in the fossil record. But while all scientists agree that evolution happens, there are two competing schools of thought regarding the pattern and speed of evolution.

The **gradualist** school sees species changing in small steps through time by slow directional change within a lineage, producing a long graded series of differing forms.

The **punctuated equilibria** school is based on a model in which species are relatively stable and long-lived in geological time, and that new species appear during outbursts of rapid speciation, followed by the differential success of certain of the newly formed species.

Most practising biologists see no problem in assuming that evolution is probably a mixture of these two forms.

The age of the earth.

> The world was created on 22nd October, 4004 BC at 6 o'clock in the evening.
> —James Ussher, Archbishop of Armagh (1581–1656), *Chronologia Sacra*. (Daniel Boorstin, *The Discoverers*, p. 451, has October 26 at 9 am).

We now take it for granted that the Earth is 4.6 billion years old, that life is about 3.8 billion years old, multicellular life is about 580 million years old, hominids are around 5 million years old, and humans are generally thought to be no more than 2.5 million years old, with modern humans having been around for maybe one or two hundred thousand years, give or take a few tens of thousands of years.

Working out the age of things always begins with finding a clock, but when you get down to it, almost anything can work as a clock. Let's take the early attempts at working out the age of the Earth.

At first, the problem was one between the biologists and the physicists. Way back in 1753, a French naturalist, Georges Louis Leclerc, Comte de Buffon (1707–1788) saw that the horse and donkey were similar and suggested that asses were just horses which had degenerated over time, as the Earth cooled.

Large and "more perfect" ancient mammoths and mastodons were replaced by their degenerate descendants, the elephants, and so on. Clearly, Buffon was on the way to an idea of evolution. He also said "In inland districts, on mountain peaks and in places farthest from the sea, shells, skeletons of sea-fish and marine plants are found, which are just the same as the shells, fish and plants now living in the sea, which are, indeed, exactly the same."

The earth, he decided just had to be older, so he carried out cooling experiments with large hot iron balls. From his data, Buffon deduced that the planet was at white heat, 75,000 years ago, and had carried life for only 40,000 years, which was no help at all to geologists and biologists.

Other French scientists like Jean-Baptiste Joseph, Baron de Fourier (1768–1830) looked at the Earth's central heat, proven by higher temperatures in mines and volcanic activity and assumed in a similar way that the whole Earth was once very hot, and that the temperature of the Earth was now falling.

Since they knew how fast things cool down (very fast, when there is no continuing source of heat), the physicists could set an upper limit on the Earth's age, just by taking its temperature. Buffon's cooling-Earth model continued to attract scientists, right through the 19th century. Fourier was supported strongly by people like Lord Kelvin.

In 1862, Kelvin estimated the age of the Earth, from its cooling time to be between 20 and 400 million years. Again, he assumed no internal heat sources. Kelvin was not alone in his beliefs, as this anonymous *Scientific American* comment from 1892 shows, though now the point of debate was how old the Sun was:

> Sir Robert Ball places the day when the world will come to an end, as we know it, about four or five million years distant. The heat which he estimates that the sun originally contained would supply its radiation for about 18,000,000 years at the present rate. It is believed that the sun has already dissipated four-fifths of the energy with which it may originally have been endowed, and this brings us to the conclusion that it will last 5,000,000 years longer.

The problem was that while the physicists would later use radiometric dating to provide a far greater age for the Earth, that was in the future. In the 19th century, they knew nothing of solar fusion, and had no idea where the Sun got its energy from.

All they could see was that no known procedure could provide energy at that sort of level for more than a few million years. If they assumed the Sun was just radiating heat and so slowly cooling down, they could show that the Earth would have been too hot for life in the recent past when the Sun was hotter.

The 19th century saw many discussions about the source of the Sun's energy. One theory was that it came from the release of gravitational energy when the sun's material contracted. In this case, however, the calculated life expectancy of the sun was, in our eyes, rather short.

Then slowly, scientists began to emphasise the huge time that life had been around before humans. When they suggest the lengths of time involved, they prefer to speak by analogy. If life had been on Earth for 24 hours we are told, humans have been around for just a few minutes.

Or, taking the old definition of a yard (from the tip of the king's nose to the tip of the king's finger), we were told that a single stroke of the nail file on the king's longest finger would wipe out the entire human presence in the history of life. Mark Twain suggested that human history was like the coat of paint on the very tip of the Eiffel Tower.

And then there was the problem that quite recently, the Earth's surface must have been too hot to walk on. Somewhere, they realised, they were making a wrong assumption.

Still, there were other ways to estimate the planet's age. Jumping back for a moment, Edmond Halley (1656–1742) invented a way of estimating the age of the planet from the oceans. He suggested measuring the salinity of the sea, and then finding the rate at which salt was added to the sea each year.

> Now if this be the true reason for the saltness of these lakes, 'tis not improbable but that the ocean itself is become salt from the same cause, and we are thereby furnished with an argument for estimating the duration of all things, from an observation of the increment of saltness in their waters.
> —Edmond Halley, 'An Attempt to find the Age of the World by the Saltness of the Sea', *Philosophical Transactions*, Volume 5, 218, 1749.

This method always gives low estimates because it does not allow for losses of salt to the land, either as salt spray, as halite (rock salt) deposits or as subduction losses. Still, it was a start, though it took a while for anybody to try Halley's method, but the Irish physicist John Joly calculated in 1899 that the sodium content of the oceans was 1.5×10^{16} tons, yielding an age of 97 million years.

The data were a little unreliable: the same method used today would give an age of about 250 million years. Then came the night in 1904 when Ernest Rutherford, a rising star of physics, was to give the Bakerian lecture at the Royal Institution in London:

> I came into the room which was half dark, and presently spotted Lord Kelvin in the audience and realized that I was in for trouble at the last part of the speech dealing with the age of the Earth. To my relief, Kelvin fell fast asleep, but as I came to the important point, I saw the old bird sit up, open an eye and cock a baleful glance at me! Then a sudden inspiration came, and I said Lord Kelvin had limited the age of the Earth, provided no new source of heat was discovered. That prophetic utterance refers to what we are now considering tonight, radium! Behold! the old boy beamed upon me.

Still, Kelvin wrote a letter in 1906, later published in the *British Weekly*, where, speaking of the Sun and the Earth, he said "It seems almost infinitely improbable that radium adds practically to their energy for the emission of heat and light". At that time, 'radium' meant not only the element, but also radioactivity in general.

Another solution was to measure the rate at which sediment was being deposited, find the total thickness of sediments, and use this to estimate the Earth's age. In 1860, this method had yielded an age of three million years, but by 1910, the same method returned an age of 1.6 billion years!

The problem is that there are times when a given location is gathering no sediment, or is even being eroded. To get a proper estimate, we need to measure the thickest sediment bed laid down in each and every period and sub-period.

On this basis, for age estimation purposes, the Earth's total sediment deposition can be taken as 150 kilometres thick: at a sedimentation rate of 30 cm per thousand years, that makes the oldest fossil-bearing sediments about 500 million years old. This, of course, cannot really allow for the effects of erosion, and takes no account of sediments converted to metamorphic rocks, and so lost to the record.

In 1905, Ernest Rutherford and Bertram Borden Boltwood used radioactive decay to measure the age of rocks and came up with a date of 500 million years, assuming that helium was the end-product of radio-active decay. Two years later, Boltwood took the metal lead as the end product, and estimated the Earth's age as 1.64 billion years.

In 1913, Rutherford and Joly looked at radioactive decay in minerals and calculated an age for the planet that was more like 400 million years. Over the years, the age estimates grew, slowly, but even an age of 90 million years meant that some other method was needed to fuel the Sun, and as we have seen, other estimates set the minimum age at 1.6 billion years. Something was wrong!

In 1920, an experiment showed that a helium atom has less mass than four hydrogen atoms. The British astrophysicist Sir Arthur Eddington realized that nuclear reactions in which hydrogen was transformed into helium might be the basis of the Sun's energy supply, using Albert Einstein's formula $E=mc^2$.

The mass difference between the four hydrogen atoms at the start, and the helium atom that results, is accounted for because the mass has been converted to energy—and that is where $E=mc^2$ comes in, but it would be some years more before we understood about nuclear fission in heavy elements and nuclear fusion in light elements.

By 1931, on the basis of assorted radioactivity and geological data, the age of the earth was taken to be at least two billion years. In the end, Hans Bethe would provide the answer to the way the Sun kept going, showing that it would last far longer than people had thought. Then the debate died off as the physicists developed better models for dating rocks, and as they developed more complex models for the Earth's heat engine.

In 1954, a revised estimate, based on the best information, put the earth at 5 to 6 billion years. Today, we know of rocks that are nudging the four billion year mark, and we are fairly confident in setting a 4.6 bya date of birth for our planet.

And now, it's time to look at how we estimate the ages of things.

Dating things: technicalities.

Dating takes two forms: it can deliver an absolute age in years or an age relative to other events. Relative dating is sometimes all that is available to us. All geological dating methods come with a small amount of uncertainty, because they rely on probabilities and inference, based on the best available data.

The rocks of the Earth and other things can be dated in a variety of ways, and the method(s) we use will depend on whether we are dating a Viking village in northern England, early human remains from Australia, a hominin from Africa, or sediments in New Zealand.

Some dating methods can be interfered with by contamination of the sample, but applying several methods can help avoid the risk of error from this source. The oldest fossil traces we know of go back to about 3800 million years, but as most rocks of that age have been long since destroyed, life may be a little older.

Thermoluminescence can identify how long some things have been buried. The thermoluminescence clock is 'reset' when the objects are exposed to direct sunlight.

Isotope dating works with many igneous rocks, and this can be used to determine absolute limits to the age range of fossils lying between two igneous layers.

Ice cores provide good evidence of past climates and temperatures. The cores preserve stable isotope ratios in water and gases, and solids like volcanic ash.

And most importantly, given the assumptions of stratigraphic correlation, we can often fix a date range for a rock, based on the beds above and below it. Two of the established methods of dating are of little use to geologists, so there is little here on carbon or tree ring dating.

Thermoluminescence works because crystalline materials store energy when they are hit by ionising radiation from natural radioactivity. This energy is lost when pottery and flint are heated above 400°C, or when sediments are exposed to sunlight as they are deposited.

Once covered, and with the "clock set", the crystals slowly begin to gather energy again. The light which is emitted on testing will then be proportional to the stored energy, and that in turn is proportional to the amount of time that the material has been buried, or left unheated.

The sample needs to have a surface layer 2 mm thick stripped off all over, and must then have a minimum size of 1 cm x 2 cm x 3 cm. In addition, a sample of at least 250g of the surrounding soil is needed for proper testing. The soil sediments need to be such that they would have been exposed to sunlight just before they were buried.

Pottery older than about 200 years can be dated, while burnt flint can be dated in the range 10,000 to 300,000 years, and sediments can be dated in the range 1000 to 300,000 years. The error bands range from 7% to 12%.

Luminescence dating is especially useful when there are no suitable organic deposits, or when there is organic material available, but the relationship between the organic materials and the archaeological contexts is uncertain. This method also reaches back well beyond the period in which carbon dating is useful, a mere 50,000 years or so, but the best aspect is that the date is calculated for the artefact, rather than for the associated materials.

The only extra condition is that the site either has to be tested for local radiation levels, or extra soil samples need to be taken from an area 30 cm around the artefact, so the local radiation levels can be determined later.

Alpha-recoil-track dating can be used on Quaternary (recent) micas. This involves looking at the damage caused by alpha particles striking mica and causing damage to the crystal lattice in young volcanic rocks which are hard to identify any other way. Since the damage can only have happened since the rock formed, and since the sources nearby can be identified and assessed, this offers a way of dating the rocks. In practice, the split surfaces are etched with acid. The tracks are less resistant to acid attack, and after being etched, they show up more clearly.

Radiometric dating works by looking at the proportions of source atoms and "daughter" atoms in a rock sample. For example, as zircon crystals form in cooling magma, they capture radioactive uranium, but no lead, so there is no lead in the crystals when the "clock" starts.

Uranium decays into lead, so when you study volcanic rock formations, you know that the lead atoms you find in zircon crystals came from the decay of uranium (refer to the list of isotopes and their half-lives, below).

Comparing the proportion of lead-207 to the proportion of uranium-235 in zircons can tell geologists how long ago the magma solidified into rock. Then, if you have a lava bed that is just above a fossil bed, you have a latest possible date for the fossil, and if there is another lava bed below, you have an earliest possible date for the fossil as well.

Depending on the age of the rocks being studied, other element pairs may be used instead. Potassium-argon dating works because when a mineral such as *orthoclase* ($KAlSi_3O_8$) crystallizes from a melt, it contains none of the inert gas argon. So any argon-40 (^{40}Ar) in the sample is the decay product of potassium-40 (^{40}K). To date a mineral sample, we only need to heat the sample in a vacuum to release the ^{40}Ar and then measure the released gas in a mass spectrometer.

If we know the concentration of ^{40}Ar, the current concentration of potassium and the $^{40}K/^{39}K$ ratio, we can calculate how many half-lives must have elapsed since the time of crystallization. Metamorphic events can involve rocks being melted, and letting any gases escape. In this way, the clock can be reset by releasing all or part of the argon as the orthoclase recrystallises.

In the case of regional metamorphism, the metamorphic rock will be completely surrounded by other metamorphic rocks, but the contact metamorphism produced in a small area by volcanic intrusions and flows can be a potential problem.

Isotopes and their half-lives

Carbon–14 changes to nitrogen–14 with a half-life of 5.715×10^3 (5715) years

Uranium-235 changes to lead-207 with a half-life of 7.04×10^8 (704 million) years

Potassium-40 changes to argon-40 with a half-life of 1.26×10^9 (1.26 billion) years

Uranium-238 changes to lead-206 with a half-life of 4.46×10^9 (4.46 billion) years

Thorium-232 changes to lead-208 with a half-life of 1.4×10^{10} (14 billion) years

Rubidium-87 changes to strontium-87 with a half-life of 4.88×10^{10} (48.8 billion) years.

We shall never get people whose time is money to take much interest in atoms.
— Samuel Butler (1835 - 1902), *Notebooks*.

And that brings us to the interaction between those busy money people and the future of the planet.

15. End-of-the-earth science.

Two views of climate change: a widespread (and regrettably anonymous) skewering of denialists; and Jens Galschiot's installation 'Unbearable' in Copenhagen. The curve shows the rise in CO_2 levels.

King tide, January 2018, next to Sydney Opera House, invading the Royal Botanic Gardens. King tide, January 2018, over-topping an extended sea wall, North Harbour, Sydney.

Why climate matters

We live on the surface of the Earth, we are made of Earth stuff, and for the past 2 million years, our technology has been based on the use of Earth stuff. More to the point, our future is intricately and desperately tied up with the future of a planet that we have been treating very roughly.

It is now more essential than ever that humans understand how the Earth works—not just the ways that volcanoes, earthquakes, tsunamis, the rock and water cycles operate, but the practical effects of mining and pursuing wealth, where we get our water from, how climate systems work and interact.

We need to know the origins of the Earth (and the solar system), how we found out about it, how we came to understand the rock and water cycles, how the landscape has been shaped and will be shaped in the future.

We need to understand plate tectonics, which explains so much; how folds and bends happen in rocks; what drives volcanoes, geysers and thermal pools;

how rocks and crystals are formed; how dykes and columnar basalt form; the nature of pumice, granite, sandstone, shale, metamorphic and other rocks.

In a world where weather extremes are becoming more common as the climate changes, we need to understand floods, mudslides, erosion and rock falls; we need to understand the way tides and waves interact with the shore; how rocks weather and decay, giving way without warning; and we need insights into the planet's past, a form of understanding most easily shaped by a close look at fossils.

Environmental educators say there are three things we can do to bring about environmental awareness: we can get people to *know* about the problems, we can get them to *care* about the problems, and we can get them to *take action*: if we achieve any two of these, the third will follow.

In 1997, just 98 of 171 Nobel laureates signed a statement making these points:

- Global warming is under way and our overuse of fossil fuels is partly to blame.

- Climate change is projected to raise sea levels; increase the likelihood of more intense rainfall, floods, and droughts; and endanger human health by greater exposure to heat waves and encroachment of tropical diseases to higher latitudes.

- Climate change is likely to exacerbate food shortages and spread undernutrition by adversely affecting water supplies, soil conditions, temperature tolerances, and growing seasons.

- Climate change will accelerate the appalling pace at which species are now disappearing, especially in vulnerable ecosystems. Possibly one-third of all species may be lost before the end of the next century.

- Continued destruction of forests will undermine the environment's natural ability to store carbon, thereby enhancing global warming.

A month earlier, on 25 July, Bill Clinton was a great deal more certain, when he addressed several hundred invited guests at the White House:

"We see the train coming, but most ordinary Americans in their day-to-day lives can't hear the whistle blowing."

It was a message echoed by one of the speakers, John Holdren, a professor of environmental science and public policy at Harvard University, who commented on the many linkages between the developed world and the developing world in a time of climate change: "you can't sink just one end of a boat".

The nature of science and the way science is done means that nothing is ever totally absolutely proved to be true. Even so, there are many things that all scientists believe and accept, because to believe otherwise seems ludicrous.

The precautionary principle is commonly applied when there is clear but insufficient evidence. The consensus of climate scientists is that the evidence of a catastrophic threat is clearly sufficient, but if it were not, it is sensible to act as though it is. There's a famous Joel Pett cartoon where somebody asks "What if it's a hoax, and we create a better world for nothing?"

The precautionary principle is a melding of "look before you leap" and "He who hesitates is lost". It requires informed judgement by scientists, not the greedy squawling of con-men seeking a quick buck.

This leaves us with the problem of what to believe when two competing viewpoints both seem feasible. This is the position at the moment with global warming, the greenhouse effect and rising sea levels, because the small changes we have seen *might* be caused by something other than the greenhouse effect. Perhaps the world would have warmed up like this anyhow, maybe the warming is part of a natural cycle, or perhaps we are seeing what we expect to see. Pigs may also fly. In short, one of the competing viewpoints is vanishingly infeasible.

By 1999, the people who spoke of climate change were no longer the mavericks, crazy outsiders: it was the deniers who were beyond the fringe. The fraction of Nobel laureates supporting that statement today would be far higher.

In 2015, one Nobel laureate (physics, 1973) made a confused speech about climate change being a good thing. In response, 36 younger laureates signed the 2015 Mainau declaration, stating in part: "We believe that the nations of the world must take the opportunity at the United Nations climate change conference in Paris in 2015 to take decisive action to limit future global emissions." The world order changes…

Australian laureate, Peter Doherty, explained why a few ancient scientists take such an odd view: "You're an eminent, long-retired scientist, and you take a controversial opinion because otherwise you'd be ignored. It's very common." The perceptive reader may note that the long-retired person is not named here.

How climate is changed.

Infrared (IR) active gases, mainly water vapour (H_2O), carbon dioxide (CO_2), and ozone (O_3), are naturally present in the Earth's atmosphere. These gases all absorb thermal IR radiation emitted by the Earth's surface and atmosphere. Our atmosphere is warmed by this mechanism and, in turn, emits IR radiation, with a significant portion of this energy acting to warm the planet's surface and the lower atmosphere.

As a result, the average surface air temperature of the Earth is about 30°C higher than it would be without atmospheric absorption and reradiation of IR energy. If we removed the greenhouse gases, we would see permanent ice almost everywhere outside of the tropics, and dry conditions in the tropics themselves, with all the water be tied up as ice. So the "greenhouse effect" is not bad or worrying on its own: the worrying thing is the runaway effect which seems to be following on from our adding extra greenhouse gases, through our industries.

Not all of the greenhouse gases have the same power: water vapour probably contributes 60% of the warming power, so why are we worried about carbon dioxide and methane? The simple answer is that we can do something about some of the other greenhouse gases. Water vapour doesn't *control* the temperature, it is controlled *by* the temperature. The first four columns in the table below are adapted from the Fourth IPCC Assessment Report (AR4) released in 2007, the last three are from AR5, from 2014.

Global Warming Potential and Atmospheric Lifetime for Major Greenhouse Gases

Greenhouse gas	Relative power	Atmospheric Lifetime (years)	Sources	Pre-industrial concentration (ppb)	2011 concentration (ppb)
Carbon dioxide, CO_2	1	100	Fossil fuel combustion; Deforestation; Cement production.	278,000	390,000
Methane, CH_4	25	12	Fossil fuel production; Agriculture; Landfills.	722	1,803
Nitrous oxide, N_2O	298	114	Fertilizer application; Fossil fuel and biomass combustion; Industrial processes.	271	324
Chlorofluoro-carbon 12, CCl_2F_2	10900	100	Refrigerants.	0	0.527
Hydrofluoro-carbon 23 CHF_3	14800	270	Refrigerants.	0	0.024
Sulfur hexafluoride SF_6	22800	3200	Electricity transmission.	0	0.0073
Nitrogen trifluoride NF_3	17200	740	Semiconductor manufacturing	0	0.00086

And that's the science.

A history of climate change.

We have known about what we used to call 'global warming' for quite a while. What is different now is that most reputable atmospheric scientists believe human activity is driving the modern warming of our climate. All the same, now we know that global warming is a bad description, we call it 'climate change'. Under any name, it's the same beast, and the same looming disaster.

Nor is the suspicion that humans are to blame as recent as the critics and denialists would have you believe. The problem used to be that there was not a lot of hard science in the arguments, which come down to logic, reason, careful modelling—and interpretation that is likely to be biased by a generous serving of self-interest among the nay-sayers.

That has changed in the last ten years, as we saw how the climate spin-doctors were using the same crooked tactics that were used to hide the harm that tobacco does. Nowadays, nobody denies that the Earth is getting warmer, because the evidence is there.

In December 2019, Australia recorded its six hottest days ever, but the trend was apparent in 1950, when George Kimble reported in *Scientific American* that the northern limit of wheat-growing in Canada had moved northward some 200 or 300 miles (call it 400 kilometres), adding that farmers in southern Ontario were experimenting with growing cotton.

While the Canadian cotton industry seems not to have taken off, he reported another trend that continues to this day, the northward retreat of the permafrost:

> In parts of Siberia the southern boundary of permanently frozen ground is receding poleward several dozen yards per annum.

The matter open to question back then was the cause. Kimble noted that the Domesday Book listed 38 vineyards in England in 1086, in addition to those of the Crown. He pointed also to the Greenland colony which was frozen out, back around the mid-1400s and other evidence that climates change.

He also looked at Biblical evidence on the distribution of date palms to suggest that conditions in 1950 were much like those of Biblical times, providing a picture of a climate that fluctuates around a mean.

Now about the 'greenhouse effect': in cold climates, a greenhouse is a glass shed which allows sunlight to shine in, where much of the energy is absorbed and changed to heat. Glass is less transparent to heat than it is to light, but a greenhouse does not just trap warmth that way: it also holds a body of warm air around the plants, and protects them from wind-driven evaporation. So while we still speak of 'greenhouse gases', it is rare to hear anybody mention the *greenhouse effect* these days.

Way back in the 1820s, Joseph Fourier realised that heat-trapping might occur. Then in 1856, an American scientist, Eunice Newton Foote, identified carbon dioxide as the most likely threat, before John Tyndall said much the same thing in 1861. (Tyndall usually gets the credit, but Eunice Foote was first!)

Then Svante Arrhenius reminded us in 1896 that both water vapour and carbon dioxide were 'greenhouse gases' (escaping that bad analogy is hard) and so water and carbon dioxide would play a role in making the planet get warmer.

He also considered changes that might be happening, and consulted Arvid Högbom, who just happened to know all about carbon dioxide sources and sinks. Carbon dioxide was coming from animals when they breathed, from volcanoes, and from humans burning fossil and other fuels.

Arrhenius thought the human additions were a very small part of the total in the air already, perhaps one part in a thousand was added by the burning of coal, and there were probably checks and balances. He estimated that it would take 3000 years to double the atmospheric levels of carbon dioxide, but that such a doubling would raise world average temperatures by 5 to 6°C.

In 1896, the CO_2 level was around 290 parts per million: in 2016, the value was estimated at 396 parts per million: we had travelled one third of the distance in 120 years. In 2018, it was 407.4 ppm, and in May 2019, it reached a seasonal peak of 414.7 ppm. My back-of-the-envelope scribblings suggest we will double the 1896 value by 2090, in just under 200 years, rather than 3000 years.

To Europeans back in the 1890s, the warming effect seemed nothing to worry about, because nobody had stopped to consider the cascades, the flow-ons that might be driven by that rise in temperature. A thermodynamics expert, Walter Nernst, even wondered if it would be feasible to set fire to uneconomical and low-grade coal seams, so as to release enough carbon dioxide to warm the Earth's climate deliberately!

Scientists are slow to move to a new model, a new way of understanding, something called a paradigm, and as mentioned earlier, I lived through the plate tectonics paradigm shift. There was fuss and bother along the way, but in the end, the good science was recognised and accepted.

In the 1990s, global warming was in much the same position, with scientists arguing furiously, even when they agreed on the main principles, and as in the puzzle of the wandering continents, the key evidence is all there. The problem is that once again we are stuck with a bad analogy, just as the early 1960s saw us hung up on "continental drift".

That aside, the cost of disagreement and bickering is remarkably different. It mattered not at all if people disagreed about plate tectonics (except, perhaps, that it makes tsunamis like the 2004 Indian Ocean tsunami easier to understand), but global warming will be a major disaster for humanity, and any delay has the potential to cost lives. To understand this, we have to accept some puzzling propositions.

To take one example, the formation of cold salty water in the Norwegian Sea is probably what stops Dublin and New York being iced-in each winter. This is because of the cold brine that drives a current known as the Conveyor, which in turn drives the Gulf Stream. The Gulf Stream takes warm water from the Caribbean and swirls it up around the North Atlantic, contributing to fogs and breaking icebergs loose, but keeping many ports warm and open, even in winter.

Just as the prion proteins of mad cow disease have more than one stable form, so do weather patterns, and if the weather once drops into a new pattern, we may not be able to bounce it back to where it started.

A golf ball in a wok lies at the bottom, and if you move it and let it go, it will roll back down. That is a stable system. A golf ball, sitting on a long cardboard tube doesn't fall, so we might say it is stable, but if you knocked it over, it wouldn't come back to this position. We say it is metastable.

The golf ball on the tube is stable to small nudges, but only within limits. Humpty Dumpty had two positions, one on the wall and one off it, and according to the nursery rhyme, the second was a position of no return. On the wall, Humpty Dumpty was metastable, but beside the wall, he was stable, and broken.

Climate patterns are either stable or metastable. If they are pushed too hard, they may 'flip' into a new metastable pattern (or even break), and only then, too late, do we discover that they were metastable (or even breakable). The best example of a probably metastable pattern is the monsoon system that waters much of Asia and the north of Australia, but El Niño and Indian Ocean Dipole are other possibles.

Climate scientists worry that severe changes may deliver a push that will take a metastable pattern away from what we know, and there might be no way of returning to the original pattern. The good news is that as northern Europe freezes over, the glaciers which are now melting away fast will be replenished, lowering sea levels. The increased snow cover will also increase the reflectivity of the northern hemisphere, and that may cool the planet down a little. We just have to hope it does not trigger a new stable pattern that happens to be an ice age.

The actual changes that might follow any breaking point are hard to predict. They are unlikely to be spectacular and major, and probably they will do their harm stealthily, when roads, bridges, port facilities and cities are flooded, or when agricultural land is lost, either by being covered by the sea or as a result of drastically changed rainfall patterns.

If rock is exposed in Antarctica, this could lead to a low pressure zone over the icy continent that could change weather patterns around the world. It hasn't happened yet, but we need to learn from history. Ten years ago, no politician would take a long-term view and force the changes needed in the next thirty to forty years, when most of them are elected for a mere three to four years, before they face the voters again.

It is easier to bleat plaintively that there is no real agreement among the scientists yet (there is, actually), or that some eminent scientists (they aren't eminent: just look at where their funding comes from) believe there are other explanations. That saves the politicians from having to act—and the honesty of scientists in saying that they cannot be sure just how things will go wrong allows devious short-term opportunists to prattle that "the scientists aren't sure…".

Politics is a marvellous human discovery. It is a pity that politicians still have to discover humanity and consider its prospects. It is likely that politics, dithering, duck-shoving and shilly-shallying will make this disaster happen, and many of the effects will seem to be unrelated to the climate.

Take dengue (pronounced den-GAY) fever, which is caused by the dengue fever virus, which is transmitted by the *Aedes aegypti* mosquito. The geographic range of *Aedes aegypti* is limited by freezing temperatures that kill overwintering larvae and eggs, so dengue virus transmission is limited to tropical and subtropical regions.

Aedes albopictus is also capable of spreading dengue fever. As a rule, the *Aedes* mosquitoes are recognizable by their striped legs (which give *Aedes albopictus* its nickname of 'tiger mosquito'), and the fact that they bite by day.

Dengue fever involves an internal haemorrhage that sometimes leads to shock—a drop in blood pressure and failure of blood cells to meet the metabolic demands of the body. It is a leading cause of death among children in Southeast Asia, killing about 1% of all cases. It includes four distinct viruses or serotypes, dengue 1 through dengue 4. As in the case of malaria, mosquitoes become infected with dengue after taking a blood meal from a dengue-infected person.

People infected with dengue virus develop dengue fever or dengue haemorrhagic fever. Dengue fever is also known as 'breakbone disease' because of severe headache and joint pain associated with it. Dengue haemorrhagic fever is far more serious than the rarely fatal dengue fever.

After a short incubation period of 1 or 2 weeks, the mosquito can transmit the infection to a susceptible person. An infection with any of the four serotypes confers protective lifelong immunity, but only to that serotype. The risk of developing haemorrhagic dengue appears to be increased among people later infected with a different serotype. In recent years, haemorrhagic dengue has become increasingly common in tropical America.

Climate change is expected not only to increase the range of the mosquito but would also reduce the size of the mosquito's larval size and, ultimately, its adult size. Since smaller adults must feed more frequently to develop their eggs, warmer temperatures would boost the frequency of double feeding and increase the chance of transmission, which will happen when the first person bitten is carrying the virus.

Warmer temperatures reduce the incubation time for the virus. The incubation period of the dengue type-2 virus is 12 days at 30°C, but seven days at 32 to 35°C. Half the world's population is currently at risk from the disease, and it has recently become a serious problem in Latin America. Brazil alone had a quarter of a million cases in 1997.

Dengue is hard to eradicate once it is established. In Australia, it reappeared in north Queensland in 1981 after being absent for some 25 years, and spreads each year through the areas of northern Australia where the *Aedes aegypti* mosquito is found, though cases are reported from across Australia each year, as a result of people being infected in either the north of the continent or overseas.

There have been suggestions in the past that global warming could lead to a spread of the *Aedes aegypti* mosquito, and thus the disease, but at the moment, Australian cases seem to be limited to about 200 a year. There is, however, a massive increase in cases across the whole of the western Pacific.

How cracking continents made snowball Earth.

In 2004, Yannick Donnadieu and colleagues described a model that *may* explain the Earth turning into a gigantic snowball 750 million years ago. It seems that the break-up of ancient continents could have been responsible for this global disaster, which hinges on the planet needing just the right amount of carbon dioxide. Too much is a problem, as we are beginning to see, but too little is equally a problem.

The researchers used a computer model to show how this movement of land masses could have removed large amounts of the greenhouse gas carbon dioxide from the atmosphere. Geologists believe that during at least two periods between 550 and 800 million years ago, ice sheets may have reached all the way from the poles to the Equator, a situation nicknamed 'snowball Earth'. This could only happen if there was very little CO_2 in the atmosphere.

There was probably an earlier snowball phase that ended at about the time the Yarrabubba crater was formed between Sandstone and Meekatharra in Western Australia. This crater appears to be the world's oldest known asteroid site, and in early 2020, researchers from Curtin University's School of Earth and Planetary Sciences, analysed the minerals zircon and monazite that were 'shock recrystallized' by the asteroid strike, at the base of the eroded crater.

From this, they found that the Yarrabubba crater was formed 2.229 bya. This, as we have seen from chapter 1 was a time when emerging plants were taking up all the carbon dioxide, making the planet colder, so perhaps the impact somehow reversed the snowball effect. We know that in the 400 million years after the impact left a 70 km crater in Australia, there were no glacial deposits formed, anywhere in the world.

For now, that remains a conjecture, but as we know, atmospheric CO_2 acts as a blanket, stopping heat from the Earth's surface escaping into space. Exactly how this CO_2 could have been removed from the atmosphere, causing global cooling, is unknown. The researchers claim that when one large continent broke into smaller fragments, this led to an increase in rain over the land. More rain caused more weathering of the rocks in these land masses, which absorbed large amounts of CO_2 from the atmosphere.

And now for something completely different. Climate change deniers say oil comes from living things, so burning 'fossil fuel' is natural: but what if it turned out that oil is really a mineral, not the result of dead life-forms being processed? It's an unlikely hypothesis, but if the fruit loops want to "maintain the debate", why not send down a few wrong'ns, to mess with their tiny minds?

A theory about oil.

Some fuels, like peat, coal, and perhaps oil may be derived from the fossilised remains of plants and animals. Standard wisdom says all the oil and coal that we find is organic, and so must have originated with organisms.

This is testable in some cases: we can certainly find plenty of fossils in coal, confirming that coal was formed when dead plant and occasional animal matter was buried in a swamp under the right conditions. We can see peat, brown coal, black coal and anthracite, and we can show that these are always found in sedimentary rock. We call these energy sources fossil fuels because we regard them as a form of buried solar energy, fossilised sunshine.

Every so often, a scientist comes up with what sounds like a totally crackpot idea. That is, in terms of what other scientists believe, it is a crackpot idea. Alfred Wegener wanted people to accept the idea of continents moving, and people dismissed him as an eccentric or a fool. Louis de Broglie made the crazy suggestion that electrons might really be waves, and almost failed to get his doctor's degree because of it.

Wegener died without recognition, though his theory of continental drift (which we now know in an amended form as plate tectonics) is standard stuff in your textbooks. Louis de Broglie was luckier, because Albert Einstein heard about his strange idea, and suggested gently that de Broglie might in fact be correct, and de Broglie lived to see the electron microscope (which treats electrons as waves) become a standard laboratory tool.

Wegener's case is a bit more typical, for few 'crackpots' get an easy time of it. More than that, most of the crackpot ideas turn out to be wrong. Yet without those strange ideas, science would never grow. Thomas Gold had to comfort himself with that thought, each time a geologist sneered at his ideas about where oil comes from. That, and the knowledge that scientists can change their minds.

Scientists usually work with a particular paradigm until evidence arises to make the old paradigm unacceptable. There have been many failed paradigm shifts, because scientists are only swayed by the evidence. When the scientists proposing a change are as astute and capable as the late Thomas Gold was, people need to ask themselves what evidence they should look for, either to support or refute the paradigm shift that Gold offered.

Gold was a famous physicist, one of three astronomers who worked out the steady-state theory of the universe, which has now been replaced by the Big Bang theory of the origin of the universe. He lived to see that theory overthrown, and now he was attacking an older, and more deeply accepted theory. He could not accept that our world's hydrocarbons are biogenic, made by living things.

When we first discovered petroleum, said Gold, it was close to the planet's surface, and chemists then thought that the only place you found carbon chemicals was in living things. They even named carbon chemistry organic chemistry, because it was the chemistry of organisms. Oil was made of organic chemicals, so obviously it had to come from organisms.

Now we know that comets contain 'organic' chemicals, and so does Jupiter. Nobody argues that the methane on Jupiter came from giant Jovians breaking wind, and nobody assumes there are little green people all over the comets, producing the organic stuff there. If we were to discover oil today, said Gold, we would never be so silly as to claim that it came from plants and animals, not with the knowledge we have now of other bodies in the solar system.

The geologists sneered at this. How much oil has been found in igneous rock? they asked. Gold accepted this question cheerfully. Not a lot, he said, because geologists are set in their ways, and they only drill for oil in sedimentary rock, where the oil sometimes gets trapped as it rises to the surface. He had, he claimed, extracted 12 tonnes of hydrocarbons from granite in Sweden, most of it coming from dolerite veins that have intruded into the granite from below. The veins either weakened the granite, or carried the hydrocarbon with them, he said.

The Arabian Gulf oil fields, according to Gold, have no common features at any depth, except that they are over an area of great seismic activity. This area contains 60% of the world's recoverable hydrocarbons. From the mountains of south-eastern Turkey down to the Persian Gulf, the plains of Saudi Arabia and the mountains of Iran, there is a continuous band of oil-fields, but nobody can find an adequate supply of source rocks to account for the oil that is there.

There is simply no 'coherent geology' beneath the surface to explain why the oil is found there, he said. The rocks are of all types and all ages, with nothing in common. But they are all rich in oils, and the oils are chemically identifiable, right through the area. They must have a common origin, said Gold, but some of the rocks are fifty million years younger, and were formed when the climate, the biology, everything in the area had changed. According to Gold, there is just no way the oil could have come from the rocks that have formed since life evolved.

In other places as well, we find oil provinces that stretch much further than any surface geological features. The only thing that is common is the deep seismic activity.

Then we come to Gold's other problem: where did the living things that supposedly formed the oil get their carbon? If they got it from carbon dioxide in the air, through photosynthesis, there could not have been enough for life to keep going. So, said Gold, there must have been a continuous supply of carbon compounds for life to keep going. On his calculations, the earth's atmospheric CO_2 must have been replaced 2,000 times in the past 500 million years.

The source of our hydrocarbons, he suggested, is about 150 km below the surface, seeping upwards when it can. Look at Indonesia, he said, where the movement of the Australian plate is causing activity below the surface, and there are huge oilfields. Look at California, where two plates are separating. Look at the match-up between seismic activity and oilfields in the rest of the world, he said.

It was true, he said that we often find petroleum in sedimentary rocks, but that, he said, was merely because we have a paradigm that says that we should look in sedimentary rocks, and so we only drill oil wells in sedimentary structures.

We were trapped in a 19th century paradigm, he said, one that held, until well after Friedrich Wöhler synthesised urea and William Perkin synthesised the first organic dyes in 1856, a paradigm that is reflected in the very name of the science that Perkin initiated, organic chemistry.

Back in the 19th century, as people began to drill for oil and use it, they naturally assumed the carbon compounds were organic, formed from living things. Even Pluto has hydrocarbons, but where did Pluto's methane come from? There are no swamps or cows on Pluto, yet there is methane there. These organic chemicals come from a distinctly non-organic background.

Just for now, the oil companies have not been rushing to take up exploration leases on the world's granite belts. In the future, we might just see a paradigm shift that leads them to do so, but even then, the oil would still be fossilised sunshine in a sense, for all of the solar system's other hydrocarbons must have had their origin inside the sun, or some other earlier star, and the stored energy in them is derived from a star's nuclear furnaces.

That leaves me wondering about the Yarrabubba asteroid: might it have smashed into a large deposit of inorganic oil? The best answer: more research is needed. Science often says that.

> The author was fortunate enough to interview Thomas Gold when he was in Sydney in the 1990s, to attend a public debate that he took part in, and then to dine with him and a number of Sydney physicists in a dinner that was unforgettably punctuated by the delivery of a fatogram to somebody at the next table.
>
> No, 'fatogram' is not explained in the glossary. Look it up.

16. The end.

Science is a great many things…but in the end they all return to this: science is the acceptance of what works and the rejection of what does not. That needs more courage than we might think. It needs more courage than we have ever found when we have faced our worldly problems.
— Jacob Bronowski, *The Common Sense of Science*, 148.

It is better to debate a question without settling it than to settle a question without debating it.
— Joseph Joubert (1754 - 1824)

Do you hear the children weeping, O my brothers,
Ere the sorrow comes with years?
— Elizabeth Barrett Browning (1806 - 1861), *The Cry of the Children*.

We are wealthy and wasteful but this can't go on. If we don't eat dog biscuits, we could end up eating our dog instead.
— Magnus Pyke (1908–1992)

Entropy is Time's arrow
— Sir Arthur Eddington

From ghoulies and ghosties and long-leggety beasties,
And things that go bump in the night,
Good Lord, deliver us!
— Anonymous, Cornish prayer.

Death and Famine and War and Pollution continued biking towards Tadfield. And Grievous Bodily Harm, Cruelty To Animals, Things Not Working Properly Even After You've Given Them A Good Thumping but secretly No Alcohol Lager, and Really Cool People travelled with them.
— Neil Gaiman, Good Omens: *The Nice and Accurate Prophecies of Agnes Nutter, Witch*.

This is the way the world ends
This is the way the world ends
This is the way the world ends
Not with a bang but a whimper.
—T. S. Eliot, *The Hollow Men*.

There may be a choice.

In the short term, we either freeze or fry, while in the long term, we decay. Still, the problems are not due to start just yet. Over the next 7 billion years, our Sun will age, gradually exhausting its fuel supply and collapsing into a white dwarf, but before it does this, it will mushroom in size, shining so brightly that it will fry the surfaces of all the inner planets, including the Earth.

So we don't have 7 billion years, just 3.5 billion, give or take a few million years, before the biosphere is sterilised, swept clean of all life forms as we know them. There is some good news though. First, life is only about half-way through its time on Earth, and almost anything could happen in the intervening period.

Aside from any strange evolutionary pathways that we could not even begin to guess about, we may be "rescued" by a close encounter between our solar system and a passing star. The quotation marks are necessary above, because this sort of rescue is very much of the frying pan/fire variety.

From computer modelling, we know Jupiter is vulnerable to gravitational interactions with a passing star, that its orbit could be disrupted quite easily by a star passing by, close to our Sun.

This in turn could affect the Earth, with even a modest change in Jupiter's orbit having the potential to bring about catastrophic changes to the Earth. There is perhaps one chance in 100,000 that the Earth could be flung into the Sun in the next 3.5 billion years, but the Earth might also be thrown out into deep space, where it would take about a million years for the oceans to freeze solid, while life of sorts would continue near hydrothermal vents on the ocean floor, which are warmed by radioactive heat from deep within the Earth.

In other words, life on Earth might become similar to the life that scientists hope one day to find on Europa, the moon of Jupiter which has thick ice sheets that seem to cover liquid oceans, deep below the surface.

Long after our solar system has faded away, our galaxy will move into the Degenerate Era (which sounds far more ominous than "a degenerate era"). By then, the only remaining starry things will be white dwarfs, brown dwarfs, neutron stars and black holes.

Over time, the available dark matter will be consumed, and then the mass of white dwarfs and neutron stars will begin to smear out through a process called proton decay, then the enormous mass of the black holes that will have formed must eventually break into thermal radiation, photons and other decay products.

It gets even gloomier, because once the black holes have radiated away, there will just be a diffuse sea of electrons, positrons, neutrinos and radiation suspended in nearly complete and total blackness.

We will no longer be able to read books, so hurry up and finish reading this.

Disaster theories.

Popular fictional science is full of disaster scenarios, because those tales sell well. Sometimes, though, the scenarios are popular, but not science. Let us consider the view that asteroid strikes make volcanoes erupt.

The world's flood basalt provinces came from the largest eruptions of lava on Earth, with known volumes of individual lava flows exceeding 2000 cubic kilometres. By comparison, the ongoing eruption of Kilauea volcano on Hawaii produced just 1.5 cubic kilometres from 1968 to 2016, and a 2018 eruption of 0.25 cubic kilometres was regarded as "huge".

The very largest flood basalts are the Deccan Traps and the Siberian Traps (as we saw in chapter 1, 'trap' comes from a Sanskrit word meaning 'step'). The name is used because of the way the layers of basalt weather and erode later to produce stepped hillsides). The Columbia River flows shown below are rather smaller than the Deccan traps, but the same stepped pattern is visible there.

Stepped flood basalt flows, Columbia River, USA.

A number of the flood basalts seem to have formed at times close to the occurrence of certain extinction events, in particular the Newark outpouring of a million cubic kilometres, some 201 million years ago; the Deccan outpouring of about 2 million cubic kilometres, around 66 million years ago; and the Siberian one, also of some 2 million cubic kilometres, around 249 million years ago.

The Deccan outpour lies close to the Cretaceous-Tertiary boundary, formed at the time when the dinosaurs all died, and the Siberian event matches closely the Permian-Triassic boundary, while the Newark event matches the end of the Triassic.

The probability of having three major volcanic events that would each typically last about a million years should occur within 1 million years of major extinction events during the last 250 Myr (of which there are about 12) is estimated at about one in ten thousand.

This has tempted many in the past to assume that these volcanic outbursts were responsible for the extinction events, and when an asteroid in Mexico was associated with the Cretaceous-Tertiary extinctions, some vulcanologists wondered if the impact of the asteroid might have triggered basaltic flows.

How serious would such an event be? The only flood basalt eruption since written history began was the 1783-84 eruption of Laki in Iceland. This produced a basaltic lava flow of 565 cubic kilometres, which represents only 1% of the volume of a typical large igneous province (or LIP) flow, but the eruption's environmental impact resulted in the deaths of 75% of Iceland's livestock and 25% of its human population died from starvation. If such a relatively small eruption happened today, all air traffic over the North Atlantic would probably be halted for three to six months.

So it seems possible that an eruption bigger than that would possibly be enough to trigger an extinction event, but all the same, the idea that volcanoes can erupt when the Earth is smacked by a large comet or meteorite has become a popular idea among disaster enthusiasts. That may be so, but it seems there is no proof to back the claim up, and science is not a popularity contest.

Not only is there no firm evidence that an impact started a volcanic eruption on Earth or on any other planet, there is no known mechanism by which this can occur. It seems that the idea was helped along by some bad science.

In the pre-Apollo days of space exploration, astronomers noted the common occurrence of dark material, usually assumed to be lava in impact basins on the Earth side of the moon. They thought that impacts caused lava to upwell in the biggest craters after they had formed, eventually filling them.

Good photos of the lunar far side, taken by the Russian probe Zond 3 in 1965 showed that the far side basins were not filled with basalt. Later samples returned from the moon by the Apollo missions showed that the mare basalts are considerably younger (up to about 1 Gyr) than the basins in which they lie.

That aside, there seems to be no way that the impact of an asteroid could punch a deep enough hole to let all that basalt out. Even the 100-km diameter Chicxulub crater barely disturbed the Moho beneath it, with less than a few km uplift beneath the centre. Under these circumstances pressure relief melting seems very unlikely, even in the largest known terrestrial craters.

So, exciting as the scenario may be to movie makers, it seems to be an idea without legs, and that isn't all the bad news for those who enjoy a bit of doom and gloom. Other precious theories fail to stand up, as well.

We will, of course be polite and not mention the possible benefits we may have got from the Yarrabubba event, if it got us out of a Snowball Earth.

Asteroids.

People used to believe that asteroids were all the remains of a broken-up planet, but now they are more likely to be considered as parts of a planet that never formed. In 1980, Luis Alvarez, his son Walter, with Frank Asaro, and Helen Michel, argued that the dinosaurs had been wiped out by the impact of some sort of asteroid or comet, landing on the Yucatan peninsula in Mexico.

This happened 65 million years ago, and enriched the iridium in the K-T layer, which is geologist-speak for the boundary between Cretaceous deposits and Tertiary ones. Iridium has also been found at the Permian-Triassic boundary, though not at the concentration found in sediments from the time of the dinosaur extinction

Asteroids are irregularly shaped rocky bodies, lying between the orbits of Mars and Jupiter, or those same bodies if they have been deflected from that region. If or when a huge asteroid hits the Earth, the catastrophic destruction it causes and even the impact winter that follows might be only a prelude to a different, but very deadly, phase that starts later on: an ultraviolet spring.

First would come the enormous devastation, huge tidal waves, and a global dust cloud that would block the sun and choke the planet in icy, winter-like conditions for months, as spelt out in certain recent popular movies. The asteroid impact would be of a magnitude similar to the one that occurred around the Cretaceous-Tertiary boundary, when there was a massive extinction of many animals, including the dinosaurs.

That one is believed to have hit, just off the Yucatan Peninsula, with a force of almost one trillion megatons. After the immediate effects already listed, the atmosphere would become loaded with nitric oxide, causing massive amounts of acid rain. As a result of the acid, lakes and rivers would have reduced amounts of dissolved organic molecules, which would allow a much greater penetration of ultraviolet light.

At first, dust clouds would keep the ultraviolet out, but this just sets the stage for a greater disaster later on, since many animals depend on some exposure to ultraviolet light to keep up their biological protective mechanisms against it. In the absence of any UV, these protective systems would cease to operate, so the animals would be unprotected when the ultraviolet spring came.

It is likely the dust cloud would shield the Earth from ultraviolet light for just over a year, by which time UV levels would be as they are now, as the dust settled. Then UV levels would continue rising until they were at least double current levels, about 600 days after impact.

The levels of ultraviolet-related DNA damage would be about 1000 times higher than normal, while general ultraviolet damage to plants would be about 500 times higher than normal. The Yucatan asteroid hit a portion of the Earth's crust that was rich in anhydrite rocks, raising a 12-year sulfate haze that blocked much of the ultraviolet radiation.

Anhydrite rocks cover less than 1% of the planet's surface, so we may not be so lucky next time. The collision will be devastating, and the impact winter deadly, but the ultraviolet spring will finish off the survivors.

The threat will probably come from an Apollo asteroid, one of a group of asteroids whose orbits cross that of the Earth. They are named after the first of their kind, Apollo, which was discovered in 1932 and then lost until 1973. They are so small and faint that they are difficult to see except when they are close to Earth, and they can come close from time to time.

In January 1991 the Apollo asteroid 1991 BA passed 170,000 km from the Earth, the closest observed approach of any asteroid, and the Yucatan impact may have involved another Apollo asteroid.

Sooner or later, an asteroid will splash down in an ocean. Asteroid 1950 DA is 1 kilometre across, and there is a slight chance of it hitting Earth in 2880, but there are probably others, still to be spotted. An asteroid that size would blast a cavity about 18 km across, all the way down to the sea floor. As the ocean rushes in, waves will radiate out at 800 km/h, and rather than the one big wave that movie-makers like to show, there will be a number of them, starting small and getting larger, one every three or four minutes.

This variation is important if you are ever caught in a tsunami zone. The wave system also leads to a 'drawing-down', where the sea appears to go out: if you see this, run uphill, but it will probably only put off the inevitable!

The dinosaur event was nothing…

Things do occasionally go wrong. First there was the death-of-the-dinosaurs asteroid, then we learned that Earth's most severe mass extinction, an event 250 million years ago which wiped out 90% of all marine species and 70% of land vertebrates, was triggered by a collision with a large space object—either an asteroid or a comet.

This does not mean that the object actually struck and killed each of those life forms, but rather, it triggered a series of events, such as massive volcanism (perhaps) and changes in ocean oxygen, sea level and climate. Those changes in turn led to species extinction on a wholesale level.

The story captured headlines around the world when it broke in *Science* in 2001, as any good disaster story will, and the evidence still looks fairly solid. So far, the site of the impact, somewhere on the single giant continent of Pangea, remains unknown, but it left traces that can be seen today, and this is a good lesson on How Scientists Find Out.

The traces are in the form of buckminsterfullerenes, or buckyballs, with two "noble" (meaning chemically nonreactive) gases helium and argon trapped inside their cage structures. Fullerenes, which contain at least 60 carbon atoms, have a structure resembling a soccer ball or a geodesic dome. The domes and the molecules are named for Buckminster Fuller, who invented the dome.

The giveaway lies not in the buckyballs themselves, because complex carbon molecules could have formed in other ways, but in the noble gases trapped inside, which have an unusual ratio of isotopes. Where terrestrial helium is mostly helium-4 with only a small amount of helium-3, most extra-terrestrial helium is helium-3, and that is the isotope showing up inside the buckyballs.

Things like this form in carbon stars. The extreme temperatures and gas pressures in carbon stars seem to be the only way extra-terrestrial noble gases could be forced inside a fullerene. It is likely that the gas-laden fullerenes were formed outside the Solar System, and their concentration which has now been detected at the Permian-Triassic boundary means they were delivered by a comet or asteroid.

The researchers estimated the comet or asteroid was 6 to 12 kilometres across, or about the size of the asteroid that left the huge Chicxulub crater near what is now the town of Progresso on Mexico's Yucatan Peninsula 65 million years ago.

But how do you estimate the size of an object after all this time? Researchers relied on two factors: if the body were smaller than 6 kilometres the effects wouldn't be seen globally, as they appear to have been; if it were larger than 12 kilometres there would have to be more gas-laden fullerenes distributed globally.

The fullerenes, which contain helium and argon, were extracted from sites in Japan, China and Hungary, where the sedimentary layer at the boundary between the Permian and Triassic periods has been exposed.

The evidence was not as strong from the Hungarian site, possibly because the sample came from slightly above or below the boundary layer, but the Chinese and Japanese samples offered strong evidence. Fullerenes are found at very low concentrations above and below the boundary layer, but they are found in unusually high concentrations at the time of the extinction.

The mass extinctions at the Permian-Triassic boundary have long been known. For example, the 15,000 species of trilobite, do not appear above the boundary. The extinction appears to have taken place in a time span of between 8000 and 100,000 years—a remarkably short time in geological terms.

But while the extinction was obvious, scientists expected strong evidence of the element iridium at the boundary, of the sort seen at the Cretaceous-Tertiary boundary in the dinosaur extinction event. Iridium has in fact been found at the Permian-Triassic boundary, but not nearly at the concentration found in sediments from the time of the dinosaur extinction.

The difference might simply be down to the two space bodies that slammed into Earth having different compositions. While the findings remind us that impact with large space bodies can be bad for life on Earth, there is also evidence an earlier collision might have been the key to life starting here in the first place.

Science is like that.

17. Glossary.

> The difference between a piece of stone and an atom is that an atom is highly organised, whereas the stone is not. The atom is a pattern, and the molecule is a pattern, and the crystal is a pattern; but the stone, although it is made up of these patterns, is just a mere confusion.
> —Aldous Huxley (1894 – 1963), *Time Must Have a Stop*. 1945, chapter 14.

Science takes the ideas and theories and names, but instead of making a mess of them, it puts them into more elaborate notions that usually need *more* new names.

ablation: Strictly, this means all the loss processes from a glacier: melting, evaporation, sublimation, wind erosion, and calving, anything that removes snow or ice from a glacier or snowfield, but the word is also used in a restricted sense, where the ablation zone is the front of a glacier. See chapter 3.

aeolian sandstone: A sandstone formed from windblown deposits, and laid down dry, rather than in a watery environment. See chapter 5.

angle of rest: Also known as the angle of repose, this is the maximum angle that a set of particles will remain piled at, after avalanching has just taken place. While the angle of repose is mainly of interest to geologists studying the nature of sand dunes and of cross-bedding in sandstones, the angle of repose is also relevant to the study of sloping talus or scree deposits at the bases of cliffs, and to explaining the way in which an ant lion catches its prey. An understanding of the angle of repose will also help to explain why snow fields are sometimes buried under avalanches of snow from higher slopes.

The angle of repose depends on the material, with both the shape of the grains and the surface properties of the grains being important. Smooth glass beads have a measured angle of repose of 23.9 degrees, while spherical quartz grains deliver an angle of repose of 31.5 degrees in air and 31.4 degrees in water.

A sand with coarse elongated grains, like *Lithothamnium* sand, can have an angle of repose of 36.2 degrees, bettered only by long-grain rice, at 36.4 degrees. If smooth glass beads are etched with hydrofluoric acid to roughen their surfaces, the angle of repose can increase by as much as 10 degrees, showing that there is a frictional effect involved.

In most cases, the materials will hold a much steeper slope, called the angle of initial yield. Spherical quartz sand grains, for example, can pile up to 45.1 degrees in air, and 44.6 degrees in water, but these slopes are unstable. Sooner or later, the surface will tumble down in an avalanche, and settle at the angle of repose. This suggests an interesting home enquiry or science project, collecting sand from various places, and measuring the *angle of initial yield* and *angle of repose* for each sand. The difference between these two angles is called the *angle of dilatation*. Typical published values for this are around 8 to 13 degrees. You would probably need to relate this to the shape of the sand grains, and maybe the amount of salt, organic matter or shell grit in the sand. See chapter 5.

aquifer: Porous and permeable beds or rocks. Aquifers hold much of the world's water supply, which may be exploited directly by sinking wells or by pumping water from the aquifer into a reservoir. The Dakota sandstone (USA) and the chalk in the London Basin (UK) are important aquifers, as are the geological structures which make Australia's huge artesian basins. See chapter 9.

basalt: One of the igneous rocks. A black to medium-grey aphanitic rock. Most basalts are fairly uniform, though they may contain phenocrysts of plagioclase and olivine. Basalt is the world's most abundant form of lava, mostly found in flows, with smaller amounts found in igneous intrusions. See chapter 1.

basin: A large low-lying area, in which sediments are deposited.

biotite: Chemically this mica mineral is a complex silicate of potassium, iron, aluminium and magnesium. It is also known as black mica. It is black, brown or dark green, and common in igneous and metamorphic rocks.

Bowen Reaction Series: How minerals are deposited from hot magma. Ultramafic peridotite rocks come out first (at the highest temperature), then basaltic rocks, then andesites, and last of all, the granitic rocks. By an odd set of circumstances darker rocks come out before lighter ones, as the magma cools from 1400°C to 800°C. If you want more, silly you, but you now have all the words you need. Here's a picture.

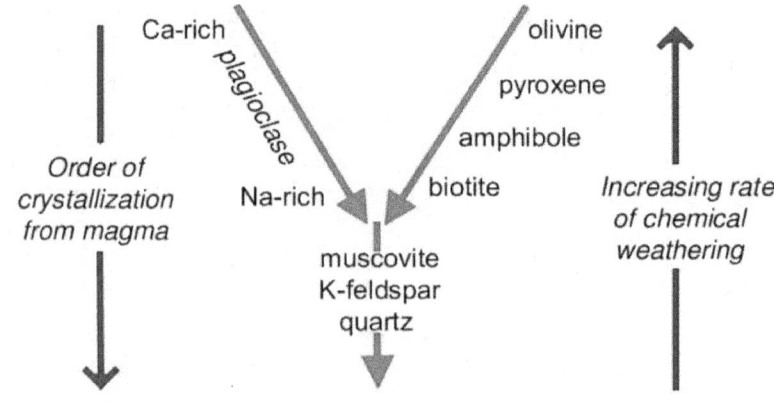

A diagram of the Bowen Reaction Series.

columnar jointing: Vertical joints that form when igneous rock cools, producing a pattern of vertical 'pipes', usually with about seven or eight sides. It is common in basalt flows. See chapter 1.

conglomerate: A sedimentary rock composed of quartz and rock fragments, formed when gravel is consolidated. See chapter 5.

contact metamorphism: The effect when a dyke, sill or flow has caused local heating of the pre-existing rocks with which the hot rock has made contact. Unlike regional metamorphism, this can happen at shallow depths, operates over small distances, and usually does not involve pressure as well. See chapter 6.

cross bedding: Also known as current bedding or cross-stratified bedding, this is a geological structure which is common in sandstones laid down in conditions where the sand has been pushed forward into place, rather than by being dropped into place. When this happens, each bed is laid down at an angle matching the angle of rest of the sand under those conditions.

Cross bedding often arises in aeolian (wind-blown) deposits, where sand is pushed over a crest and tumbles down a slope, or in delta deposits and river beds, where sand is pushed along and then tumbles down the front face of an advancing sand bank. See chapter 5.

disconformity: Parallel layered strata which are separated by an irregular surface of erosion, which has an appreciable relief—in other words, a local gap in the geological record that is produced when rocks are eroded away, and then new sediments are laid down on top. Compare this with an unconformity, and note that disconformities are sometimes seen as no more than a subset of the unconformities. See chapter 12.

dyke: Also seen as **dike**, a sheet-like body of intrusive igneous rock which cuts across the bedding or structural planes of the host rocks, unlike a sill, which pushes its way in between beds. Under field conditions, the dyke is usually either more or less resistant to weathering than the surrounding rock, and so may be visible either as a wall or a ditch. See chapter 1.

dynamic equilibrium: If I have crystals of salt sitting in a saturated brine solution, some of the chloride and sodium ions in solution may attach to the crystals, but on average, just as many ions will leave the crystals. We say the solid and the solution are in dynamic equilibrium.

erosion: Wearing away, the removal of the land surface by water, ice, wind or other agencies. This is often confused with weathering, which is the breakdown of rock material which is later eroded. See chapter 4.

erratic: A rock fragment transported into an area from outside, found either incorporated into the sediment or lying free. It is sometimes called a glacial erratic. See chapter 3.

fault: A plane of weakness in the Earth's crust, along which blocks of rock may move. See chapter 5.

feldspar: The name given to plagioclase and orthoclase.

fossil: The remains or traces of a once-living organism, usually referring to organisms that lived before the last Ice Age. The word literally means 'buried' or 'dug up', but now it has a narrower meaning. The art of finding, preparing and interpreting fossils is a very challenging one. A fossil may be either a body fossil (such as a bone or shell) or a trace fossil (such as a burrow, track or imprint). See chapter 10.

glacier: A slow-flowing river of ice.

gneiss: One of the metamorphic rocks which can be derived from granite, diorite, shale, mica schist, slate or rhyolite, among others. It is coarse-grained, with distinct layers of minerals, and feldspar is usually common. See introduction and chapter 6.

granite: This is a granular member of the igneous rocks, with 20 to 40% quartz and orthoclase, and often mica and other minerals. Some granite forms by metamorphosis, but all granites are considered to have formed at great depth. See introduction and chapter 2.

honeycomb weathering: A pattern of holes formed in rocks by weathering. See chapter 3.

hornblende: Chemically this ferromagnesian mineral is one of the amphibole group. It is dark green to black.

igneous rocks: Rocks that are formed by solidification of molten magma. Examples include basalt, diorite, granite and granodiorite. See chapters 1 and 2.

isostasy: The theory that the Earth's crust can be considered as less dense blocks floating on the denser semi-molten mantle. High mountains in this model must be regions where the crust is thickest, with deep roots extending into the mantle, and this is borne out by measurements. See chapter 5.

joint: In geology, a plane of weakness cutting across bedding planes, along which groundwater may ooze, or dykes may intrude. See chapter 3.

lava: This is magma that has reached the surface through a volcanic vent. See chapters 1 and 2.

limestone: A sedimentary rock formed mainly of calcium carbonate.

magma: Molten rock, including water and gases that may be present, formed at depth by melting. If it reaches the surface, it is then called lava. See chapters 1 and 2.

magmatic differentiation: The principle that says when magma cools, some crystals form first, changing the chemical composition of the magma. See chapter 2.

marble: One of the more commercially valuable metamorphic rocks, formed when limestone is metamorphosed.

metamorphic rocks: Rocks which have been changed by a combination of heat and pressure. Examples include gneiss, slate, marble, quartzite and schist. There are two main kinds of metamorphism, called regional metamorphism and contact metamorphism, as well as burial metamorphism. See also igneous rocks and sedimentary rocks. See chapter 6

mica: Chemically this mineral is a silicate which forms thin flakes. See biotite and muscovite.

Mohorovičić discontinuity: This lies between the crust and the mantle of the Earth, and it was discovered in 1909 by Croatian seismologist, Andrija Mohorovičić (1857–1936). It was detected by a careful analysis of shear waves from earthquakes originating at a distance of less than 800 kilometres.

(Note: there should be an inverted caret over the first letter c, and an acute over the final c in the surname, as I am trying to do here: Mohorovičić. These characters often fail to reproduce, and being hard to type on English language keyboards, do not normally appear in search strings.) Key words to look up: isostasy, isostatic and Mohole (this last one is a funny story).

moraine: The name given to piles of rock, left behind by a glacier. See chapter 3.

muscovite: Chemically this mica mineral is a complex potassium aluminium silicate. It is also called white mica or isinglass.

olivine: Chemically this mineral is magnesium iron silicate. It is olive to yellow-green, and has a conchoidal fracture.

orthoclase: Potassium feldspar, or potassium aluminium silicate. It is also one of the standards in Mohs' scale of hardness, given a hardness rating of 6. It can be distinguished from plagioclase because it lacks striations.

plagioclase: Chemically this mineral is a rock-forming silicate, in the range between sodium aluminium silicate and calcium aluminium silicate. The plagioclases are classified as albite, oligoclase, andesine, labradorite, bytownite and anorthite. Albite is pure sodium-plagioclase, anorthite is pure calcium plagioclase, and the others in the series are intermediate between these.

plate tectonics: The concept that the Earth's crust is divided into a number of rigid plates that are in motion relative to each other; the plates are formed at the mid-ocean ridges (seafloor spreading) and typically destroyed in deep-sea trenches. This differs from 'continental drift' in that the motion is explained by presumed convection currents, and it also incorporates the effects of seafloor spreading and subduction of one plate by another, while the plates are not necessarily the same as the continents. See chapter 8.

pumice: A light porous volcanic stone of mixed silicates. It is often found on sea shores because the porosity allows the stone to float.

qanat: An upward-sloping well that taps a sloping water table. See chapter 9.

quartz: Chemically, this mineral is silicon dioxide (SiO_2). It is also one of the standards in Mohs' scale of hardness, given a hardness rating of 7. It is common in granite, and is one of the most common minerals in the Earth's crust. The clear crystals are known as *rock crystal*, but it is commonly white and translucent. Semi-precious varieties such as amethyst may be coloured.

regional metamorphism: The form of metamorphism which takes place at great depth, when rocks are subjected to both heat and pressure. The scale is much greater than that seen in contact metamorphism. See chapter 6.

sand: Cohesionless sediment particles of size range 2.0-0.0625 mm. Sand is subdivided into very coarse sand (2.0–1.0 mm), coarse sand (1.0-0.5 mm), medium sand (0.5-0.25 mm), fine sand (0.25-0.125 mm) and very fine sand (0.125-0.0625 mm). See chapter 4.

sandstone: A sedimentary rock composed mainly of small grains of silicon dioxide (quartz) sand.

schist: One of the metamorphic rocks derived from basalt, andesite, gabbro, tuff, shale or rhyolite. Schists are divided into chlorite schists with chlorite, epidote and plagioclase.

seamount: An elevated area of limited extent rising 1000 m or more from the surrounding ocean floor. It is also called a guyot.

sediment: Material that settles out at the bottom of a liquid when it is still. The term is used mainly in earth sciences, but turns up also in chemistry: a precipitate is one form of sediment. In the earth sciences, though, sediment is usually regarded as particulate matter that has been transported by wind, water or ice and subsequently deposited, or that has been precipitated from water; sedimentation. See chapter 4.

sedimentary rocks: Rock formed by deposition of particulate matter transported by wind, water or ice, or by precipitation from solution in water under normal surface temperatures and pressures, or by the aggregation of inorganic material from skeletal remains. Examples include shale, limestone and sandstone. See chapter 5.

shale: A sedimentary rock composed of very fine sediment, mud clay and silt, typically less than 1/16 mm in diameter. See chapter 5.

soil: The surface weathered layers of the Earth's crust and any intermixed organic material, often including living invertebrates and microbes.

specific gravity: In chemistry, specific gravity or relative density is the ratio of the weight of a given volume of a substance to the weight of an equal volume of water.

stratum: Strata (the plural form) are layers of rock set down at some past time, on the earth's surface, later to be covered by newer strata. Because of the way in which they are laid down, geologists can be confident that the age of the lowest rocks is greater than that of rocks further up in the succession, as defined in the Law of Superposition.

unconformity: The result of a gap in the geological record, where rocks have been eroded away, and later, new material has been added above. If the beds below the break are tilted, it may be referred to as an angular unconformity. If the rocks below the break are eroded plutonic or metamorphic rocks, it may be referred to as a nonconformity. If the beds above and below are parallel to each other (that is, they appear conformable), it is generally referred to as a disconformity. See chapter 12.

uniformitarianism: The doctrine that natural geological processes affecting the Earth are still operating at essentially the same rate and intensity as they have throughout geological time. It replaced the earlier theory of catastrophism. It is wrong, though, to assume that the uniformitarians said there were no catastrophes at all. This is a common misinterpretation: in fact, they argued only that there had been no giant catastrophes along the lines of Noah's flood, and that what we see could be explained by forces no greater than those to be seen operating on a daily basis. They key to their argument is that they believed the world was much older, giving slow processes the time to effect many of the changes that show up in the geological record. See chapter 14.

vesicles: A more or less spherical space in igneous rocks, formed by gas as it expands when pressure is released. Such vesicles commonly fill with mineral crystals. (The vesicles found in animals and plants are unrelated.)

weathering: Physical, chemical and biological changes in a rock, resulting from exposure to the atmosphere, that accompany soil formation from parent rock. See chapter 3.

About the author

Peter Macinnis turned to writing after his promising career as a chiaroscuro player was tragically cut short by a caravaggio crash during the *Trompe L'Oeil* endurance race. He recently did remarkably well in the early rounds of the celebrity underwater cooking program, *Moister Chef*, but he was disqualified for using dried fruits and desiccated coconut. He has a pet slug named Gladys, living in a jar on his desk, and he is an expert possum and echidna handler and ant lion wrangler. He wrote both the score and the libretto for the acclaimed opera *Manon Troppo* ('Manon Goes Mad').

His off-season hobby is composing fake CVs.

If you believed most of that, please contact me on the address below, so I can sell you a swamp. I mean, with global warming, it will soon be dry enough to build on, so get in now!

petermacinnis@ozemail.com.au

Take 2 (more reliable): Peter Macinnis is a Sydney-based science writer who writes mainly about social history, biology, mathematics, technology and colonial Australia. He has written many award-winning children's books, but he also writes for the less demanding adult market. He has been heard since 1985 on ABC Radio National, talking to adults about sciencey stuff.

He and his wife Christine travel quite a bit when not enjoying their grandchildren. Christine, who is also science-trained, is his loyal beta reader and unthanked editor. He blogs at *Old Writer on the Block*, http://oldblockwriter.blogspot.com.au/, and he is active on social media, using either his own name or the handle McManly. Some of his books are listed on the next page, and a full list of his books may be found at http://members.ozemail.com.au/~macinnis/writing/index.htm

This book is one of a large number or reclaimed and revised titles. In 2021, he persuaded or coerced lazy publishers to allow the rights to out-of-print works to revert to him. This is the last, and the details of the others are here:

http://members.ozemail.com.au/~macinnis/writing/bookshop.htm

By the same author: a small selection
National Library of Australia titles:
Curious Minds, for adults, the artists and naturalists who described Australia.
Australian Backyard Explorer, younger readers, how to map Australia, award winner.
Australian Backyard Naturalist, younger readers, Australian wildlife, award winner.
Australian Backyard Earth Scientist, Australia and its rocks, award winner.
The Big Book of Australian History, younger readers, our land from Go(ndwana) to now, four editions.
Survivor Kids, younger readers, staying alive, out-of-doors. Every foreign tourist needs this!

Puffin titles, juvenile with Jane Bowring and Kim Gamble
The Desert, a serious ecology picture book with stories.
The Rainforest, a serious ecology picture book with stories. Award winner.

Allen and Unwin titles:
Bittersweet: the story of sugar, world history, adult readers.
Rockets: Sulfur, Sputnik and Scramjets, world history, adult readers.
The Killer Bean of Calabar, world history, poisons, adult readers. (Many translations).
It's True: You eat poison every day. Juvenile, also in simplified Chinese.

Pier 9 (Murdoch Books) titles:
Australia's Pioneers, heroes and fools, adult, the true story of mapping Australia. Now in Amazon (see below).
Mr Darwin's Incredible Shrinking World, adult, world science history for the year 1859. Now in Amazon (see below).
The Lawn: a social history, adult, world technology history. Now in Amazon (see below).
The Speed of Nearly Everything, adult, worldwide. Now in Amazon (see below).
The Monster Maintenance Manual, making fun of monsters, award winner, ages 9 to 99. Now in Amazon (see below).
100 Discoveries, adult world history of science, also in German. Absorbed in *They Saw the Difference* (see below).

Black Dog title
Kokoda Track: 101 Days. My only foray into military history. Award winner, new version now in Amazon.

Five Mile Press titles: the history you didn't get in school
Not Your Usual Gold Stories: Australian history for adults: what really happened. Now Amazon.
Not Your Usual Bushrangers: Australian history for adults: what really happened. Now Amazon.

Amazon ebooks and print-on-demand, new and reprints
Australia's Pioneers, heroes and fools, adult, the true story of mapping Australia. Revised, reprinted.
Kokoda Track: 101 Days. My only foray into military history. Award winner, Revised, reprinted.
Mr Darwin's Incredible Shrinking World, adult, world science history for the year 1859. Revised, reprinted.
The Monster Maintenance Manual, making fun of monsters, award winner, ages 9 to 99. Revised, reprinted.
Curious Minds, for adults: the artists and naturalists who described Australia. Revised, reprinted.

Not Your Usual Gold Stories: Australian history for adults: revised and reprinted by the author.

Not Your Usual Bushrangers: Australian history for adults: revised and reprinted by the author.

The Lawn: a social history, adult, world technology history: revised and reprinted by the author.

The Speed of Nearly Everything, adult, worldwide: revised and reprinted by the author.

Not Your Usual Treatments, world history of strange medical tricks, adult. Amazon only.

Mistaken for Granite: earth science for rock watchers. Amazon only.

Old Grandpa's Book of Practical Poems, an anthology for grandchildren. Amazon only.

Playwiths: STEAM activities for young people at a loose end. Amazon only.

Looking at Small Things: how to find stuff to put under the hand lens or microscope. Amazon only.

They Saw the Difference: a social history of science, answering what you never asked. Amazon only.

You Missed a Bit: adult, a massive (800-page) social history of Australia. Amazon only.

Australian Backyard Explorer, younger readers, how to map Australia, award winner, now Amazon.

Australian Backyard Naturalist, younger readers, Australian wildlife, award winner, now Amazon.

For a full listing, see https://tinyurl.com/macinnisbooks, which points to http://members.ozemail.com.au/~macinnis/writing/index.htm.

www.ingramcontent.com/pod-product-compliance
Lightning Source LLC
Chambersburg PA
CBHW052342220526
45465CB00003BA/918